ULTRAWIDEBAND ANTENNAS
Design and Applications

ULTRAWIDEBAND ANTENNAS

Design and Applications

Daniel Valderas
CEIT and Tecnun, University of Navarra, Spain

Juan Ignacio Sancho
Tecnun and CEIT, University of Navarra, Spain

David Puente
Universidad Politécnica de Madrid, Spain

Cong Ling
Imperial College London, UK

Xiaodong Chen
Queen Mary, University of London, UK

Imperial College Press

Published by

Imperial College Press
57 Shelton Street
Covent Garden
London WC2H 9HE

Distributed by

World Scientific Publishing Co. Pte. Ltd.
5 Toh Tuck Link, Singapore 596224
USA office: 27 Warren Street, Suite 401-402, Hackensack, NJ 07601
UK office: 57 Shelton Street, Covent Garden, London WC2H 9HE

British Library Cataloguing-in-Publication Data
A catalogue record for this book is available from the British Library.

ULTRAWIDEBAND ANTENNAS
Design and Applications

ISBN-13 978-1-84816-491-8
ISBN-10 1-84816-491-2

Typeset by Stallion Press
Email: enquiries@stallionpress.com

Printed in Singapore.

Foreword

Traditionally, the term "ultra-wideband" (UWB) has been associated with terms such as impulse, carrier-free, baseband, time-domain, non-sinusoidal, orthogonal function and large-relative-bandwidth radio/radar. The basic feature of UWB systems is the occupancy of an extremely wide operating bandwidth, compared to conventional radios, due to their use of impulse signals. Since the late 1960s, UWB technology has been developed mainly for radio and radar systems, in particular, for government and military applications.

However, in order to use UWB technology to benefit commercial markets and concurrently protect against possible interference between UWB systems and other existing electronic systems, since 2002, several countries released the spectra of below 900 MHz and 3.1–10.6 GHz bands for the unlicensed use of UWB systems. Unlicensed commercial applications include imaging systems and vehicular radar systems, as well as communication and measurement systems. The regulated emission from UWB systems is an effective isotropic radiated power (EIRP) of less than −41.3 dBm/MHz. More importantly, UWB technology is no longer limited to impulse systems.

Such commercial UWB systems have inevitably raised unique design challenges for antennas. A great deal of effort in research and development has been expended in order to achieve the desired broadband characteristics in terms of impedance matching, group delay and radiation properties, as well as other practical requirements such as low cost and miniaturised size.

This book, *Ultrawideband Antennas*: *Design and Applications* authored by Daniel Valderas and his colleagues Xiaodong Chen, Cong Ling, Juan Ignacio Sancho and David Puente, is a timely collection of the latest progress in the antenna designs, not only for UWB, but also for broadband applications. This group is comprised of co-authors from diverse perspectives, each of whom has been very active in, and greatly contributed to, this area. Besides conventional ways to design and optimise broadband antennae, the book also addresses a systematic design process to fit specific requirements in a unique way. Therefore, if you are interested in the forefront of broadband antenna technology, this book definitely will be a welcome addition to your library.

Zhi Ning Chen
Institute for Infocomm Research
Fusionopolis, Singapore

Acknowledgements

The authors would like to express their sincere gratitude to all those who have made this project possible, which has resulted in a long and worthwhile path of effort and satisfaction. Particularly, the authors would like to thank Prof. A. Manikas (Imperial College, London) who envisaged the potential of this cooperation that has fruitfully culminated in this successful publication, and Prof. Z. N. Chen (Institute for Infocomm Research, Singapore) for starting the pages of this book with his valuable foreword.

Daniel Valderas and Juan I. Sancho are indebted to Andrés García-Alonso for his priceless encouragement. They are also grateful to César A. Rufo for the construction of the prototypes, always fabricated in a timely fashion, and to those who have collaborated with them from the very beginning, like Aitor Crespo, to those at the very end, like Raúl Álvarez. Special recognition goes to CST Microwave Studio, as a result of a fruitful collaboration, for providing the licenses that have paved the way for the development of this research work.

Daniel Valderas wishes to thank his parents, Alejandro and Isabel, for their continuous example of life … that it is an adventure more precious than writing a book. He would also like to thank his engineer, Isidoro Zorzano, who helped him with publication issues. He also acknowledges a debt of gratitude to IEEE and John Wiley & Sons, who have kindly granted permission to reprint valuable material — this has made some of the chapters in the book possible.

Cong Ling cannot thank his wife Lu Gan enough, for looking after their newborn while he concentrated on this project.

Xiaodong Chen would like to thank Lu Guo, Sheng Wang, Choo Chiau, Pengcheng Li and John Dupuy for conducting most of the research work at Queen Mary, University of London, and his colleagues in other institutes, for their courtesy in permitting the authors to quote their work.

Authors

Daniel Valderas received his M.Sc. degree and Ph.D. from Tecnun, University of Navarra, in 1998 and 2006, respectively. He is a Researcher with the Electronics and Communications Department of the Centro de Estudios e Investigaciones Técnicas de Gipuzkoa (CEIT). In 2002, he joined Tecnun and is currently a Lecturer in the Electricity, Electronics and Control Department. He was Academic Visitor at Florida Atlantic University (2003), Imperial College (2007) and Queen Mary, University of London (2008). He has participated in many different research projects on antenna fields. His research focuses on UWB and broadband antennas, RFID and MIMO antennas and antennas for body applications.

Juan Ignacio Sancho received his M.Sc. degree in Industrial Engineering from Tecnun, University of Navarra, in 1988. He worked in Telefonica S.A. and in 1990, he joined Tecnun as an Assistant Professor in Electromagnetics. In 1995, he received his Ph.D. in Electrical Engineering. He has participated in different research projects about electromagnetics and antennas. In 2006, he joined as Director of the RF Area of Tecnun. His current research interests include broadband and UWB antenna design and analysis, and RFID antennas and technology. He is currently a Professor on Electromagnetics and Antennas and Propagation in Tecnun.

David Puente received his M.Sc. degree in Electrical Engineering, specialised in Radio Communication, from Tecnun, University of Navarra, in 2007. Between 2005 and 2006, he collaborated with the Electronics and Communications Department of CEIT, thanks to a fellowship from the Spanish Ministry of Education. In 2006, he formulated his Master's Thesis about RFID Antennas Design at the High Frequency and Microwave Technology Department at Fraunhofer IIS in Erlangen, Germany. He is currently working towards his Ph.D. in Electrical Engineering at the Polytechnical University of Madrid (UPM) with a F.P.U. predoctoral fellowship from the Spanish Ministry of Education.

Cong Ling is currently a Lecturer in the Electrical and Electronic Engineering Department at Imperial College London. He received his B.Sc. and M.Sc. in Communications from Nanjing Institute of Communications Engineering, China, in 1995 and 1997, respectively, and his Ph.D. in Electrical Engineering from Nanyang Technological University, Singapore, in 2005. Before joining Imperial College, he was on the faculties of Nanjing Institute of Communications Engineering and King's College. He is currently an Associate Editor of *IEEE Transactions on Vehicular Technology*, and has also served on the programme committees of several international conferences, including IEEE Information Theory Workshop 2006, Globecom 2007, and ICC 2008. He is a member of IEEE and IET.

Xiaodong Chen received his B.Eng. in Electronic Physics from the University of Zhejiang, Hangzhou, China, in 1983, and his Ph.D. in Microwave Electronics from the University of Electronic Science and Technology of China, Chengdu, in 1988. In September 1988 he joined the Department of Electronic Engineering at King's College, University of London, as a Postdoctoral Visiting Fellow. In September 1990 he was employed by King's College London as a Research Fellow working on numerous industrial and government funded research projects. In March 1996 he was appointed to an EEV Lectureship at King's College London. In September 1999 he joined the Department of Electronic Engineering at Queen Mary and Westfield College, University of London, as a Lecturer. He was promoted to a Reader in the same college in September 2003. In October 2006, he was appointed to a full Professorship in Microwave Engineering at Queen Mary, University of London.

Contents

Chapter 1

Introduction to Ultrawideband Systems

Cong Ling
Imperial College London

In this chapter, we introduce the basics of ultrawideband (UWB) systems, the spectrum and regulations, and the advantages of UWB over conventional narrowband systems. We outline different schemes to realise UWB: impulse radio/time hopping, direct sequence, frequency hopping and orthogonal frequency division multiplexing (OFDM). Then we show the industrial standards and applications of UWB.

1.1. Overview

UWB refers to systems with very large bandwidth (Ghavami *et al.*, 2008; Aiello and Batra, 2006; Arslan *et al.*, 2006). This very large bandwidth offers several advantages including low power consumption, high date rate, high time resolution, low-cost implementation, obstacle penetration, resistance to interference, covert transmission, co-existence with narrowband systems and so on. Such advantages enable a wide range of applications of UWB to communications, radar, imaging and positioning.

Arguably, the first experiment on UWB was conducted by Hertz in 1893 (Ghavami *et al.*, 2008). In other words, the first wireless communication system was based on UWB. Hertz used spark gaps and arc discharges between carbon electrodes to generate electromagnetic waves. The usage of such wideband pulse waveforms was the dominant technique for many years after Hertz's first electromagnetic experiment. However, as technology progressed, the emphasis of communications shifted to narrowband sinusoidal waveforms. It was not until the 1990s that investigations into impulse radio sparked new interest in UWB. The technology of impulse radio was made possible by the corresponding techniques for generating short pulses developed in the 1960s.

1

Perhaps the greatest advantage of UWB is most evident from the famous Shannon formula for the capacity of a band-limited channel in Gaussian noise (Ghavami *et al.*, 2008):

$$C = W \log \left(1 + \frac{P}{WN_0} \right),$$
(1.1)

where C is the channel capacity in bits/second (bps), W is the channel bandwidth in Hz, P is the signal power in Watts and N_0 is the noise power spectral density in Watts/Hz. The Shannon formula shows that, given the noise power spectral density N_0 of the channel, the signal power P can be traded off with the bandwidth W while maintaining the same channel capacity C. In particular, we can decrease P if more bandwidth is available. On the other hand, given P the capacity C will increase with W. From an information-theoretic perspective, this trade-off between power and bandwidth motivated the development of wideband communication systems such as spread spectrum and UWB. The trade-off between power and bandwidth was known even in analog communication. Frequency modulation (FM) uses more bandwidth in return for higher signal quality than amplitude modulation (AM).

In spread-spectrum communication, the signal power is spread over a much wider bandwidth than the original (Simon *et al.*, 1994). However, the bandwidth of traditional spread-spectrum communication (on the order of MHz) is not comparable with that of UWB, which is on the order of GHz. This definition of UWB is not unique. Federal Communications Commission (FCC) in United States of America (USA) defines UWB as systems with bandwidth larger than 500 MHz or relative bandwidth W/f_c (f_c is the carrier frequency) larger than 20% (Aiello and Batra, 2006). Figure 1.1 shows the spectral mask of UWB defined by the FCC Part-15 rule and that of indoor UWB (FCC, 2002). Similar rules have been approved in other countries of the world. In particular, a very wide bandwidth of 7.5 GHz between 3.1 GHz and 10.6 GHz is available for UWB at the power emission level −41.3 dBm/MHz. In Fig. 1.1, the stricter limitation of power emission in the frequency band between 0.96 GHz and 1.61 GHz is due to existing services such as mobile communications, positioning systems and military usage.

Because of the low power spectral density of UWB, its interference can often be ignored by many existing systems occupying the same frequency bands. This property enables the unlicensed operation of UWB devices. Moreover, the short pulses of UWB lead to multipath immunity, i.e., the propagation paths can be identified due to its fine time resolution. This is because the resolution of the UWB receiver is approximately $1/W$; the larger W is, the higher is the resolution (Ghavami *et al.*, 2008). The fine time resolution is very useful for ranging and positioning. The short pulses also enable the penetration through walls and ground.

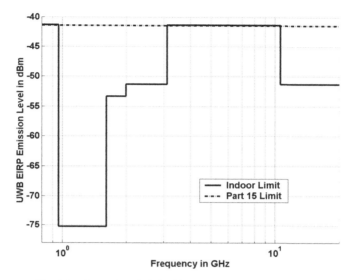

Fig. 1.1.   Spectral mask defined by FCC for UWB. (FCC, 2002.)

Moreover, pulse-based UWB does not require carrier modulation, thus reducing the size and cost of UWB devices compared to conventional narrowband systems.

1.2. UWB Schemes

Despite being seen as a breakthrough technology, UWB is closely related to existing spread-spectrum communication. In particular, impulse radio, one of the UWB schemes, is also known as time-hopping (TH) in spread-spectrum communication. Other schemes to realise UWB include direct-sequence, frequency hopping and OFDM. Quite naturally, UWB offers the same possibility of code-division multiple access (CDMA) as conventional spread spectrum does, by assigning different signature codes to different users. Here, we present an outline of different UWB schemes. The details of UWB transceiver design and signal processing can be found in Yang and Giannakis (2004) and Qiu *et al.* (2005).

1.2.1. *Impulse radio/time hopping*

Impulse radio was the first proposed UWB scheme (Win and Scholtz, 1998; Win *et al.*, 2009) and has received most of the attention in the research community. Another name for impulse radio is time-hopping. Time-hopping is not a new technology; it has been used in spread-spectrum systems for many decades. Figure 1.2 shows the principle of impulse radio, where each data

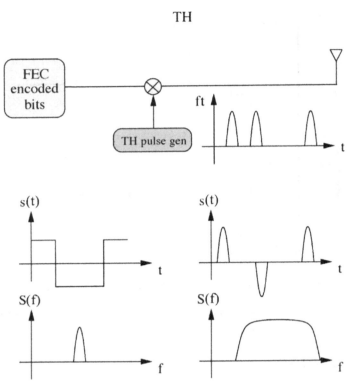

Fig. 1.2. Block diagram of TH impulse radio. $s(t)$: time-domain signal; $S(f)$: frequency-domain signal (Win *et al.*, 2009). (©2009 IEEE.)

symbol is associated to a sequence of very short duration pulses, typically of a sub-nanosecond, hence the name of impulse radio. The short pulses have their relative positions hopping in time according to a user-specific sequence. Due to the very short pulse duration, the spectrum is spread very thin from near DC to several GHz. In principle, the generation of this carrier-less signal allows for low complexity and low power-consumption transmitter implementation.

The signal emitted from the kth user in impulse radio consists of a large number of sub-nanosecond pulses (Win and Scholtz, 1998)

$$s^{(k)}(t) = \sum_{j=-\infty}^{\infty} w\left(t - jT_f - c_j^{(k)}T_c - \delta d_{\lfloor j/N_s \rfloor}^{(k)}\right), \qquad (1.2)$$

where $w(t)$ is referred to as the monocycle (pulse shape), T_f is the pulse repetition period, $\{c_j^{(k)}\}$ is the time-hopping sequence associated with user k, $c_j^{(k)}T_c$

corresponds the time shift to the jth pulse. T_c is the duration of addressable time delay bins. The data sequence $\{d_j^{(k)}\}$ changes the delay via the modulation index δ, which should be chosen to optimise the performance. N_s monocycles are modulated by a data symbol, hence the data rate $R_s = 1/N_s T_f$. When there is more than one user, the time-hopping sequences should be carefully designed to minimise the collisions. In principle, time-hopping sequences can be adapted from frequency-hopping sequences (Simon *et al.*, 1994).

A typical pulse shape for the monocycle is the Gaussian doublet, given by Ghavami *et al.* (2006)

$$
w(t) = \left[1 - 4\pi \left(\frac{t}{\tau_m} \right)^2 \right] e^{-2\pi \frac{t^2}{\tau_m^2}},
\tag{1.3}
$$

where τ_m is the parameter to determine the frequency characteristic of the Gaussian doublet. This pulse shape can be easily generated by using simple circuits and antennas. Figure 1.3 shows the waveform of the Gaussian doublet and its frequency characteristic.

1.2.2. *Direct sequence*

UWB signals can also be generated by multiplying the original data sequence with a high-rate pseudo-random sequence. The conventional technique is called direct-sequence spread spectrum. The bandwidth of the resultant direct-sequence spread-spectrum signal will be expanded by a factor as large as the period of the pseudo-random sequence. This factor is also known as the processing gain (Simon *et al.*, 1994). Obviously, to create a UWB signal, the rate of the pseudo-random sequence needs to be much much higher than the original data rate. Figure 1.4 shows the diagram of direct-sequence spread spectrum. The properties of direct-sequence spread spectrum include low probability of interception/detection, robustness to interference and multipath propagation, and CDMA.

The design of pseudo-random sequences (spreading codes) for direct-sequence spread spectrum is well documented in literature. Good spreading codes should have low auto-correlation side lobes to ensure a flat spectrum of the signal. In CDMA applications, the spreading codes should also have low cross-correlation to reduce the inter-user interference. Some classical spreading sequences are the m sequence, Gold sequence, Kasami sequence, Bent sequence, GMW sequence and the No sequence (Simon *et al.*, 1994). Other spreading sequences including chaotic sequences that offer a trade-off between auto-correlation and cross-correlation (Oppermann and Vucetic, 1997; Ling and Li, 2001).

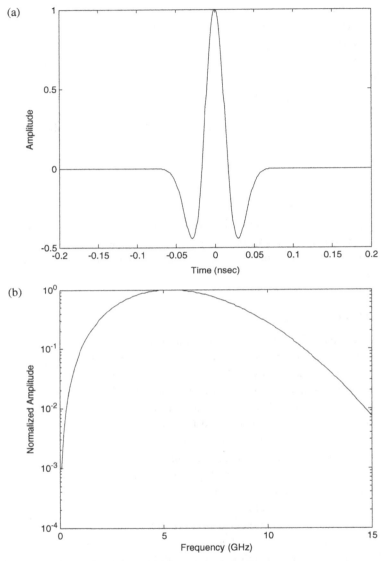

Fig. 1.3.   (a) Pulse shape for impulse radio and (b) the corresponding frequency characteristics. $\tau_m =$ 0.06 ns.

1.2.3. *Frequency hopping*

Frequency hopping is another classical approach to spread spectrum. Figure 1.5 illustrates the principle of frequency hopping, where the carrier frequency is

DS–CDMA

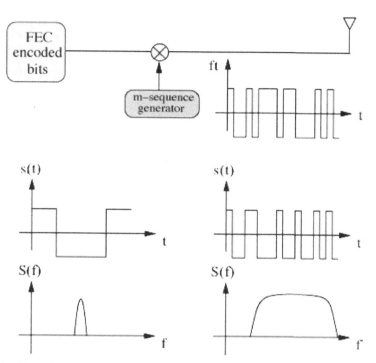

Fig. 1.4. Block diagram of direct-sequence spread spectrum. $s(t)$: time-domain signal; $S(f)$: frequency-domain signal (Win *et al.*, 2009). (©2009 IEEE.)

controlled by the frequency-hopping sequence (pattern) and hops over a certain frequency band. In a particular time constant, though, the frequency-hopping signal is a narrowband signal. The bandwidth of the frequency-hopping signal is determined by the bandwidth available, not by the rate of the frequency-hopping sequence.

Frequency hopping also has the advantages of low probability of interception, robustness to jamming and interference, and the capability of CDMA. Obviously, to maintain these properties over an ultrawide bandwidth, the carrier frequency needs to hop fast enough. Compared to the direct sequence technique, frequency hopping has another distinctive advantage in that the frequency slots do not have to be in one continuous frequency band. Therefore, in addition to spectrum flexibility, an adaptive frequency hopping system can skip those contaminated/jammed frequency slots.

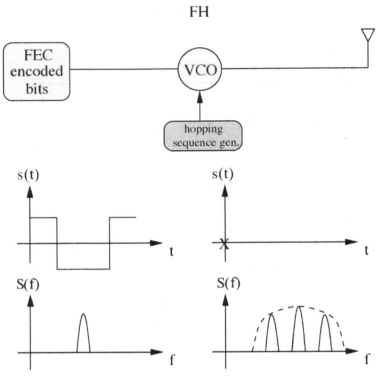

Fig. 1.5.   Block diagram of frequency hopping. $s(t)$: time-domain signal; $S(f)$: frequency-domain signal (Win *et al.*, 2009). (© 2009 IEEE.)

Techniques for the design of frequency-hopping sequences are also well known (Simon *et al.*, 1994). Here, the design criterion is the Hamming correlation, which counts the number of hits in one period of the frequency-hopping sequence. In particular, the frequency-hopping sequences should have low Hamming autocorrelation to facilitate synchronization, while in CDMA applications, they should also have low Hamming cross-correlation to reduce the number of hits between different users. In the presence of an intelligent follow-on jammer, the frequency-hopping sequences should further have long period and strong nonlinearity. Short-period or linear sequences can be easily analysed and regenerated, which likely enable the follow-on jammer to completely undermine the processing gain. For such applications, highly nonlinear frequency-hopping sequences based on the chaos approach are well-suited (Ling and Sun, 1998; Ling and Wu, 2001).

1.2.4. *OFDM*

Alternatively, an ultrawide bandwidth can be created by multi-carrier modulation with a large number of carriers. In multi-carrier modulation, the original date sequence is split into many low-rate data streams, each of them modulating different parallel carriers (subcarriers). Let $S = [S_0, S_1, \ldots, S_{N-1}]$ be a block of data symbols. The complex baseband representation of a multicarrier signal with N subcarriers is given by Han and Lee (2005).

$$s(t) = \frac{1}{\sqrt{N}} \sum_{n=0}^{N-1} S_n \cdot e^{j2\pi n \Delta f t}, \quad 0 \le t \le NT, \tag{1.4}$$

where Δf is the subcarrier spacing and NT is the data block period. In OFDM, the spacing between the subcarriers is properly chosen so that signals at different subcarriers are orthogonal to each other, i.e., $\Delta f = 1/NT$. Then, at discrete time instances $t = kT/N$, one has

$$s_k = \frac{1}{\sqrt{N}} \sum_{n=0}^{N-1} S_n \cdot e^{j2\pi nk/N}, \quad 0 \le k \le N - 1, \tag{1.5}$$

which is precisely the inverse fast Fourier transform (IFFT) of the data block. Thus, OFDM can be conveniently implemented by the IFFT; the subcarriers need not be generated physically. This significantly reduces the implementation complexity and cost.

Figure 1.6 shows the block diagram of the full digital implementation of OFDM. The data symbols undergo the serial/parallel conversion, IFFT, and parallel/serial conversion before being sent into the channel. At the receiver, the inverse operations are applied, and the data are demodulated using FFT.

Figure 1.7 shows the various subcarrier spectra overlapping with one another. Note that each subcarrier spectral peak is at the nulls of other subcarrier spectra. Figure 1.8 shows the overall OFDM spectrum by adding all subcarrier spectra together. Obviously, one can make the bandwidth very large by using a large number of subcarriers, leading to the OFDM scheme of UWB.

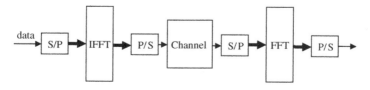

Fig. 1.6. Digital implementation of OFDM.

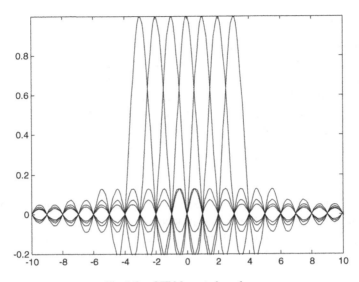

Fig. 1.7.   OFDM spectral overlap.

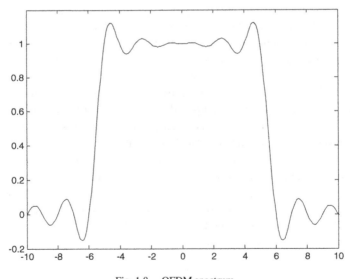

Fig. 1.8.   OFDM spectrum.

An issue with OFDM is the high peak-to-average power ratio (PAPR). As the OFDM signal is a linear combination of some random data symbols, the amplitude can vary in a large range. The central limit theorem implies that the amplitude can be well approximated by the Rayleigh distribution for a large number of subcarriers.

Han and Lee (2005) provides an overview of PAPR reduction techniques such as clipping, filtering, coding, interleaving, partial transmit sequence and selected mapping.

Another issue is the inter-carrier interference (ICI). Although the subcarriers are orthogonal to each other in principle, the Doppler shift of the wireless fading channel may cause OFDM to lose orthogonality.

1.3. Industry Standards

1.3.1. *Single band versus multiband*

Initially, UWB was envisaged for a single frequency band of several GHz. However, such a truly ultrawideband system may be hard to implement, as the design of circuits is challenging and may lead to increased cost and power consumption. Further, it is vulnerable to strong interference at sporadic frequencies.

The multiband approach (Qiu *et al.*, 2005) addresses such issues by dividing the entire UWB spectrum into several sub-bands with bandwidth approximately 500 MHz. The 500 MHz bandwidth is to satisfy the minimum instantaneous bandwidth of an UWB signal as specified by FCC. Figure 1.9 shows the multiband spectrum allocation. Obviously, any of the UWB schemes can be used within a sub-band. The multiband approach offers flexibility in allocating spectrum and dealing with interference since the spectrum does not need to be contiguous. In addition, it is easier to implement any UWB scheme over a sub-band than that over the entire bandwidth.

1.3.2. *Standards*

Within the IEEE 802.15 working group, there are several standards for wireless personal area networks. The IEEE 802.15.3 standard is an evolution of the Bluetooth specification. UWB emerged as a strong candidate. Thus the IEEE 802.15.3SGa committee was formed to define the standard. Some companies promoted the direct-sequence UWB approach, while others supported multiband OFDM. However, neither approach could obtain the necessary 75% approval and

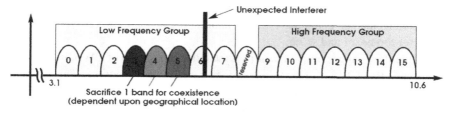

Fig. 1.9.   Multiband spectrum allocation (Pendergrass, 2003). (©2009 IEEE.)

the standard definition process became a deadlock (Aiello and Batra, 2006). Finally, IEEE 802.15.3SGa was withdrawn. Nonetheless, both approaches of UWB survive in other standards.

The IEEE 802.15.4a standard, which addresses wireless personal area networks and wireless sensor networks, has adopted the impulse radio scheme (Zhang *et al.*, 2009). The mission of IEEE 802.15.4a is to develop a new physical layer for applications such as sensor networks. The 802.15.4a standard provides an enhanced communications capability to the 802.15.4-2006 standard and also provides device ranging to enable the geolocation capability. In addition to the 850 kb/s mandatory data rate, it provides variable data rates such as 110 kb/s, 1.70 Mb/s, 6.81 Mb/s, and 27.24 Mb/s. The UWB signal bandwidth is 500 MHz for the mandatory modes of IEEE 802.15.4a, with optional bandwidths of greater than 1 GHz.

The ECMA-368/9 (WiMedia Alliance) standard for short-range high-data rate wireless connectivity uses the multiband OFDM-based UWB technology (Kumar and Buehrer, 2008). It enables data rates up to 480 Mbps. It has been adopted by USB Implementers Forum as the Wireless USB standard and by the Bluetooth Special Interest Group as the next generation Bluetooth 3.0 standard for new consumer electronics. The eventual data rates of both standards will likely be even higher. Such speeds are substantially higher than Bluetooth 2.0 (which supports a maximum speed of 2 Mb/s) and wireless LAN IEEE 802.11g (which supports a maximum speed of 54 Mb/s.) The physical layer of WiMedia defines six band groups over the entire UWB frequency spectrum (3.1–10.7 GHz). Each band group consists of three 528 MHz-wide bands. A total of ten channels are defined in each band group, which are differentiated according to the frequency-hopping patterns.

1.4. Applications

We briefly mention the applications of UWB in communications. The applications of UWB will be further explained in Chap. 10.

Short-range high-speed communication appears to be the most popular application of UWB. The industrial standards mentioned in the preceding section reflect the huge industrial interest in this application. With such standards in place, companies have begun to produce large volumes of UWB chips and systems. Applications include computers, mobile phones, digital cameras, removable hard disks, set-top boxes and so on.

While short-range communications is the killer application of UWB, it is worth mentioning that UWB is also capable of long-range communications. This should come as no surprise since the UWB technology originated in short-pulse radar systems (Aiello and Batra, 2006). Meanwhile, time hopping is a classical approach to spread spectrum, and has been used in military communications such as the Tracking and Data Relay Satellite System (TDRSS) for decades.

Another major application of UWB is ranging and localisation (Ghavami *et al.*, 2008). The precision of ranging depends on the bandwidth. The extremely large bandwidth of UWB means high precision of ranging. Specifically, the multigigahertz bandwidth of UWB leads to the ranging accuracy on the order of 1 cm. The potential of integrate ranging, location and communication makes UWB an attractive candidate for future systems.

A wireless sensor network is a network consisting of a large number of spatially distributed, cheap sensors used to monitor an environment. Wireless sensor networks have military, civil and industrial applications such as battlefield surveillance, environmental monitoring, healthcare, logistics and traffic control. The sensors usually have limited sizes, capability and battery life. Thus, UWB is a good choice of communication between the sensors because of its low-cost and low power consumption (Yang and Ginanakis, 2004). In addition, the precise localisation capability of UWB can improve positioning accuracy for wireless sensor networks. A good overview of UWB-based sensor networks can be found in Zhang *et al.* (2009).

1.5. Challenges

UWB also poses several challenges. Due to its extremely large bandwidth, the interference between UWB and narrowband systems is a major concern (Aiello and Batra, 2006). There are regulations of spectrum usage to avoid interference between different systems. The users of the spectrum have often paid a lot for it at auction, and therefore must be convinced that UWB will not cause undue interference to existing services. Moreover, all kinds of interference from existing narrowband services across the extremely large bandwidth of UWB may result in considerably stronger disturbances than background noise. This would mean that the equivalent noise power spectral density N_0 in Eq. (1.1) might be considerably larger, which may undermine the capacity promise of UWB.

The promise of low-cost implementation might also be undermined if simple circuits for signal reception and interference combating were unavailable.

Last but not least, the design of UWB antennas is considerably more challenging than conventional antennas. Conventional wideband antennas cannot transmit UWB signals without distortion. It is also more difficult to characterise UWB antennas, as traditional narrowband antenna parameters are not directly useful to UWB. The design of UWB antennas is even more challenging for small mobile terminals. These challenges will be addressed in the rest of this book.

Chapter 2

Figures of Merit for UWB Antennas

David Puente[*,a] and Daniel Valderas[†]

*Universidad Politécnica de Madrid (UPM)
†CEIT and Tecnun, University of Navarra

UWB systems are becoming increasingly important, especially due to the need for high transmission speeds. However, their range of application is not restricted to the communications field. They are also widely used in radar, collision detection systems in cars and medical imaging. In communications, the principal advantages of this technology over others are the robust characteristics it brings to multi-path environments and its low degree of interference with other systems, through the use of expanded spectrum techniques at low power.

The system's performance and characteristics are heavily dependent on the design of the radiating element. The requirements placed on UWB antennas in terms of size, phase linearity and spectral efficiency are more demanding than for narrowband antennas. Traditional parameters, such as gain and reflection coefficient, are not sufficient to analyse their performance and characteristics. It is necessary to characterise them more fully from other points of view, by adding new figures of merit.

The first part of this chapter presents the requirements, in generic terms, that are desirable in a UWB antenna. Apart from the typical demands on any radiating system in terms of matching and radiation pattern, other specifications must be guaranteed due to the requirements of pulsed transmission. Special attention is paid to modelling the pulses used in UWB, which are degraded by the radiating system, thus limiting the communication's features and performance.

---

[a]David Puente is currently working on his Ph.D. dissertation at the Universidad Politécnica de Madrid (UPM), with an F.P.U. predoctoral fellowship from the Spanish Ministry of Education.

A knowledge of the parameters used to describe these antennas, and how to apply them correctly, is essential for there to be any guarantee of success when designing them. The second part of the chapter therefore shows the figures of merit that allow different UWB antennas to be compared, and how they vary in the frequency, time and space domains.

The chapter ends with brief considerations on electromagnetic simulation in the time domain.

2.1. Requirements for a UWB Antenna

The requirements for a UWB radiator can be summarised as follows:

2.1.1. *Efficiency and matching*

In UWB, unlike in traditional communication systems, pulses of very short duration, typically tens of picoseconds, are used. Since the bandwidth is inversely proportional to the length of the pulse, the associated spectrum is very wide.

Traditionally, an antenna is understood to be broadband if its input impedance and radiation pattern do not vary significantly over at least one octave (Stutzman, 1998). A transmission system is considered to be UWB in accordance with the FCC's definition if it has a bandwidth greater than 500 MHz, or a relative bandwidth greater than 20%, defined at -10 dB (FCC, 2002). The relative bandwidth BW_r is defined as

$$BW_r = \frac{2(f_h - f_l)}{f_h + f_l},\tag{2.1}$$

where f_l is the band's lowest frequency, and f_h is the band's highest frequency.

The spectral efficiency η_{rad} evaluates the quality of matching over the whole of the frequency range. The expression that defines it for the spectrum is given in Eq. (2.2) (Chen *et al.*, 2004).

$$\eta_{rad(\%)} = \frac{\int_0^\infty P_t(\omega)(1 - |\Gamma_t(\omega)|^2)d\omega}{\int_0^\infty P_t(\omega)d\omega} \times 100\%,\tag{2.2}$$

where P_t is the power at the terminals of the transmitting antenna, and $\Gamma_t(\omega)$ is the reflection coefficient normalised to the characteristic impedance Z_o for this study.

This efficiency is a representative value for the whole spectrum, and is an expansion over frequency of the traditional narrowband parameter. Strictly, the value $(1 - |\Gamma_t(\omega)|^2)$ or reflection efficiency should be multiplied by the radiation efficiency, given by the losses in the conductor and the antenna dielectric, where applicable (Balanis, 1997). If these are assumed to be negligible, the total efficiency is given by Eq. (2.2). An obvious fundamental advantage of broadband antennas applied to UWB, in comparison to narrowband antennas, is that their total efficiency is very high, e.g., over 90% for a Rayleigh pulse.

2.1.2. Signal distortion and dispersion (ringing)

The UWB antenna deforms the transmitted signal. The antenna response to a pulse of very short duration, as is typical in UWB, is seen as a ripple after the pulse, which is called the **ringing effect**. This effect is a consequence of the antenna geometry, and translates into a frequency dispersion or time delay, which reduces the transmission speed (see Fig. 2.1).

This effect is accentuated in the case of self-similar antennas, as is shown below. Conversely, other UWB antennas, such as planar monopole antennas, have a much more constant phase centre and also use the same area of the radiating element at all frequencies. This, as is desirable, mitigates the ringing effect.

2.1.3. Stability over frequency of the transmission-reception transfer function

Stability over frequency involves evaluating the variability with frequency of the antenna's matching, gain and polarisation in the link direction, assuming that the channel conditions are constant over time. All these factors are brought together in the antenna transfer function, which compares the radiated pulse to the pulse transmitted to the antenna by the circuit.

The spectrum of the radiating system must fit the mask profile defined by the FCC, with a maximum level of $-41.3\,\mathrm{dB}_m/\mathrm{MHz}$ (Fig. 1.1). There are basically two ways to shape the spectrum of the radiated signals:

2.1.3.1. *Constant transfer function: Pulses selected directly by the source*

This technique is based on shaping the UWB pulses through an appropriate design of the circuit, in order to ensure that the spectrum of the whole radiating system satisfies the requirements. In this case, it is desirable for the antenna's transfer function to be as constant as possible over frequency, so that it does not affect the pulse.

Due to their unique properties in terms of time and spectrum, the Rayleigh family of pulses (derivatives, of different orders, of the Gaussian pulse) is widely

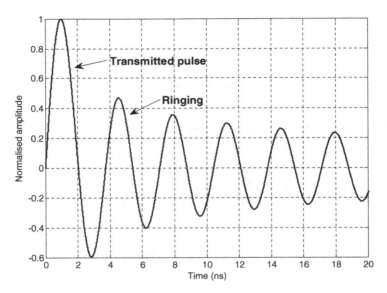

Fig. 2.1. Typical response of a UWB antenna to an input pulse.

used as signal sources for UWB systems. The corresponding equations in the time and frequency domains are Eqs. (2.3) and (2.4), respectively.

$$v_n(t) = \frac{d^n}{dt^n} \exp\left[-\left(\frac{t}{\sigma}\right)^2\right], \tag{2.3}$$

$$\tilde{v}_n(\omega) = (j\omega)^n \sigma \sqrt{\pi} \exp\left[-\left(\frac{\omega\sigma}{2}\right)^2\right], \tag{2.4}$$

where $\omega = 2\pi f$, the parameter σ is the time for which the Gaussian pulse takes the value $v_o(\sigma) = e^{-1}$ and n is the order of the derivative or Rayleigh pulse. As a general rule, the smaller the value of the constant σ, the greater the bandwidth occupied by the pulse will be. Besides, as the order of the derivative increases, the maximum moves to higher frequencies in the spectrum (Sheng *et al.*, 2003). Figure 2.2 shows the normalised Power Spectral Density (PSD) in dB for Gaussian ($n = 0$) and Rayleigh pulses corresponding to different values of n and σ.

2.1.3.2. *Variable transfer function: Concept of the antenna as a filter*

This is, in fact, the most innovative way to shape the spectrum. The antenna transfer function may help to shape the graph of the spectrum in terms of emission limits,

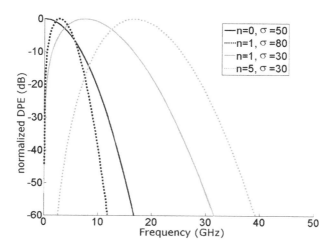

Fig. 2.2.   Normalised PSD versus frequency for different Gaussian and Rayleigh pulses.

in order to fit it to the mask. From this point of view, the antenna would behave as a filter.

2.2. UWB Antenna Parameters

Since the concept of "broadband antenna" is undoubtedly very general, it is necessary to specify the figures of merit which will be the defining test for its UWB behaviour. These indicators, which allow different antennas to be compared in terms of usefulness for UWB applications, are described below. They will be divided into three groups of parameters: those which define stability in the frequency domain, those which define it in the time domain (both in a constant propagation direction) and those which analyse variability with direction in space.

2.2.1. *Variability in the frequency domain*

An ideal UWB system would be one in which the pulse sent from the transmitting antenna, whatever the direction, would not suffer distortion on reception by the receiving antenna. The function relating the pulse to be transmitted and the pulse received, in the frequency domain, is called the **system transfer function** $H(\omega)$ (Chen *et al.*, 2004). This corresponds to the diagram in Fig. 2.3, and is defined by Eqs. (2.5) and (2.6), in which free space transmission is assumed.

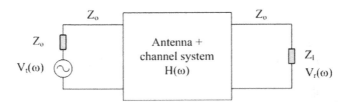

Fig. 2.3.   Schematic model of a broadband link.

$$H(\omega) = \frac{V_r(\omega)}{V_t(\omega)} = \left| \sqrt{\frac{P_r(\omega)}{P_t(\omega)} \frac{Z_l}{4Z_o}} \right| e^{-j\varphi(\omega)} = |H(\omega)| \, e^{-j\varphi(\omega)}, \qquad (2.5)$$

$$\varphi(\omega) = \varphi_t(\omega) + \varphi_r(\omega) + \frac{\omega R}{c}. \qquad (2.6)$$

In these equations:

— V_t and V_r are the corresponding signals in transmission and reception.
— ω is the angular frequency (rad/seg) at which the system operates.
— Z_l and Z_o are, respectively, the impedance seen from the terminals of the receiving antenna and the characteristic impedance of the transmission line.
— P_t and P_r are, respectively, the average power delivered to the transmitting antenna and the average output power delivered by the receiving antenna.
— R is the distance between the transmitting and receiving antennas.
— $\varphi_t(\omega)$ and $\varphi_r(\omega)$ are the phase changes introduced by the antennas, and $\varphi(\omega)$ is the phase of the system transfer function.

A pulsed transmission link does not only depend on the pulse used and the antennas considered individually, but also on their relative orientation and on the communication channel. The function $H(\omega)$ may therefore also be broken down according to Eq. (2.7) (Klemm and Tröster, 2005).

$$H(\omega) = H_{1tx}(\omega, \theta_i, \phi_i) \cdot H_{2rx}(\omega, \theta_j, \phi_j) \cdot \frac{e^{-j\beta R}}{R}, \qquad (2.7)$$

where

$$H_{1tx}(\omega) = \frac{E_{rad}(\omega, \theta_i, \phi_i)}{V_t(\omega)}, \qquad (2.8)$$

$$H_{2rx}(\omega) = \frac{V_r(\omega)}{E_{rec}(\omega, \theta_j, \phi_j)}. \qquad (2.9)$$

Equations (2.8) and (2.9) define the **transmission transfer function** H_{1tx} of antenna 1 and the **reception transfer function** H_{2rx} of antenna 2, respectively. The term $e^{-j\beta R}/R$ is the phase shift and attenuation produced by a free-space channel. E_{rad} is the component of the radiated electric field at a distance at which the receiving antenna is in the far field, and that coincides with the E_{rec} incident field.

The alignment between the antennas is defined by (θ_i, ϕ_i) at the transmitting antenna and (θ_j, ϕ_j) at the receiving antenna. The relationship between the two transfer functions, for the case when the two antennas are the same, and are oriented at the same angle, is given by Eq. (2.10) (Klemm and Tröster, 2005).

$$H_{tx}(\omega, \theta, \phi) = j\omega H_{rx}(\omega, \theta, \phi). \tag{2.10}$$

For this specific case, the problem may be simplified by only considering the transfer function for an isolated antenna. This is due to the fact that the system transfer function may be obtained from this former transfer function. The analysis can therefore focus on the properties of the isolated antennas, and thus they can also be compared.

2.2.1.1. *Magnitude of the transfer function*

The stability of the magnitude of the transfer function can be broken down by studying of the variability with frequency of the parameters included in the Friis equation (Eq. (2.11)):

$$\frac{P_r(\omega)}{P_t(\omega)} = (1 - |\Gamma_t(\omega)|^2)(1 - |\Gamma_r(\omega)|^2)$$

$$\times \ G_r(\omega)G_t(\omega)\left|\hat{\rho}_t(\omega) \cdot \hat{\rho}_r(\omega)\right|^2 \left(\frac{\lambda}{4\pi R}\right)^2. \tag{2.11}$$

Since the pulse can be expressed as a series of frequency components throughout the broadband spectrum, each of the factors in Eq. (2.11) should be as constant as possible over frequency, in order to minimise pulse distortion. Several different parameters used in studying the stability over frequency of the terms in the Friis equation are presented below.

2.2.1.1.1. Stability of the reflection coefficient

In addition to a good level of stability for the $(1 - |\Gamma_t(\omega)|^2)$ and $(1 - |\Gamma_r(\omega)|^2)$ factors, the total efficiency, given by Eq. (2.2), must be high to ensure that the level of matching is sufficient throughout the working band.

2.2.1.1.2. Polarisation stability

Polarisation coupling is defined by the factor $\left|\hat{\rho}_t(\omega) \cdot \hat{\rho}_r(\omega)\right|^2$ in the Friis equation. The aim is for it to be as close to 1 as possible, over the whole spectrum. In the case of linear polarisation, if both components are locally comparable, it is the stability of the main component that is of interest. If the axis of the antenna is the vertical z (e.g., in the case of UWB monopoles), it is the E_θ component which is of interest. If it is circularly polarised, the axial ratio must be as close as possible to 1 over the whole spectrum.

2.2.1.1.3. Gain stability and channel losses

The gain stability, $G_r(\omega)G_t(\omega)$, must be considered with respect to its absolute value and the spatial direction under consideration. It is frequently not so much the stability of the maximum gain of an antenna with the drift in the direction in which the maximum gain occurs that is of interest, but the opposite: for a plane of interest, an analysis is made of how its maximum varies with frequency, and how close it is to the other points.

Considering a simplified free space propagation model, the channel causes losses that increase over frequency, as is reflected in the $(\lambda/4\pi R)^2$ factor in Eq. (2.11). However, the power sacrificed to higher path loss is balanced by increased antenna gain for a fixed antenna aperture A_{ef}, according to Eq. (2.12) (Griffin and Durgin, 2009).

$$G = \frac{4\pi A_{ef}}{\lambda^2}. \tag{2.12}$$

2.2.1.2. *Transfer function phase: Group delay*

The stability of the magnitude of the reflection coefficient, radiation pattern, gain and polarisation do not take into account the phase difference with which the wave arrives when radiated at different frequencies. In ultra-wideband applications, all the frequency components are radiated at the same time. In order to reconstruct the pulse on reception, the stability must take into account the phase differences for the different components.

The **group delay** (*GD*) is taken as the figure of merit for studying the phase behaviour of the transfer function, and this is calculated using Eq. (2.13).

$$GD(f) = -\frac{1}{2\pi}\frac{d\varphi(f)}{df}\bigg\|_{f=f_0}, \tag{2.13}$$

where f_o is the frequency of interest.

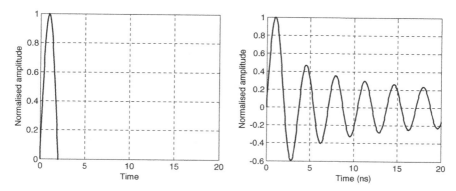

Fig. 2.4.   Example of a transmitted and received pulse in a UWB system.

An ideal UWB system would have a constant *GD*, so that the phase would vary linearly over frequency. A poor *GD* causes ringing, which considerably limits the performance of the whole communication system.

2.2.2. *Variability in the time domain: Pulse distortion parameters*

The pulse emitted by the transmitting antenna is exposed to the phenomena of distortion and dispersion. There are several figures of merit for a UWB system, which are used to analyse the degree of similarity between the original pulse and the received pulse in the time domain, (as shown in Fig. 2.4).

2.2.2.1. *Fidelity factor*

The **fidelity factor** F evaluates the degree of similarity between the shape of the pulse input to the transmitting antenna $x(t)$ and the shape of the pulse obtained at the terminals of the receiving antenna $y(t)$. This allows us to compare the shape of the pulses independently of their amplitude. It is calculated as the maximum with respect to the time delay τ of the normalised coefficient of correlation between $x(t)$ and $y(t)$, by using Eq. (2.14) (Klemm and Tröster, 2005).

$$F = \max \frac{\int_{-\infty}^{\infty} x(t) \cdot y(t - \tau) dt}{\sqrt{\int_{-\infty}^{\infty} |x(t)^2| dt \cdot \int_{-\infty}^{\infty} |y(t)^2| dt}}. \tag{2.14}$$

2.2.2.2. *Time spread*

The dispersion introduced by the UWB system shows itself in the abovementioned ringing effect on the received pulse. As a result of this, the energy transmitted is extended over time. The **99 % energy window** (E99) expresses the period of time, measured from the start of the pulse, which contains 99% of its energy.

From the value of E99 we can determine the **time spread**, which calculates the ratio between the values of E99 for $x(t)$ and $y(t)$. An ideal UWB system would have the same E99 for both the pulses, and therefore a time spread of 1.

2.2.3. *Variability in the space domain*

In pulsed transmission systems, it is not only the variability of parameters with frequency and the distortion of the pulse over time which must be considered, but also the spatial direction of the link. Variability in space is measured through statistical averages or analysis of correlation between different propagation directions, in the frequency or time domains.

2.2.3.1. *Statistical values*

2.2.3.1.1. Uniformity

The **uniformity** U (Ma and Jeng, 2005) expresses the probability that the deviation of the radiation pattern from its maximum value is less than or equal to 6 dB in the section plane under consideration, and for a given frequency. It is defined as a figure of merit by Eq. (2.15).

$$U = \frac{\text{No. of points with deviations from the maximum} \leq 6 \text{ dB}}{\text{N}^{\circ} \text{ total points in a single cut}}. \qquad (2.15)$$

In this, the maximum value is taken in the section plane for each frequency. This does not have to be the same as the maximum gain. Studies of uniformity versus frequency are useful for analysing the variability of omnidirectionality over frequency, even though the maximum for each case is different.

2.2.3.1.2. Spatially averaged transfer function (SATF)

Another way of achieving spatial analysis is based on averaging the transfer functions H_{tx} within a solid angle in space, assuming that all the directions are equally probable. Some authors therefore establish the **spatially averaged transfer function** (*SATF*) (Klemm and Tröster, 2005) as

$$SATF(\omega) = \frac{\sum_{n=1}^{N} \sum_{m=1}^{M} \overline{H}(\omega, \Delta\omega, \theta_n, \phi_m)}{NM}, \qquad (2.16)$$

where $\overline{H}$ is the transfer function normalised to the maximum amplitude within the bandwidth under consideration $\Delta\omega^b$. N and M are the numbers of propagation directions chosen in the θ and ϕ coordinates, respectively. Equation 2.17 can thus be applied, based on Eq. (2.10) in the solid angle for which the average is calculated, for a link between antennas that are identical. To obtain the system's *SATF*, Eq. (2.18) would be applied, which is based on Eq. (2.7).

$$SATF_{tx}(\omega) = j\omega SATF_{rx}(\omega), \tag{2.17}$$

$$SATF(\omega) = SATF_{tx}(\omega, \theta_i, \phi_i) \cdot SATF_{rx}(\omega, \theta_j, \phi_j) \cdot \frac{e^{-j\beta R}}{R}. \tag{2.18}$$

Each time the $SATF(\omega)$ is found, the solid angle or plane of interest, the field component and the distance for the link must be specified. By convention, the $SATF(\omega)$ is understood to be that for the antenna's transmission.

2.2.3.1.3. Spatially averaged group delay (SAGD)

Since the group delay GD is obtained from the phase of $H(\omega)$, it can be averaged over space analogously to the *SATF* so that it is more representative. This gives the Spatially Averaged Group Delay (*SAGD*), which is calculated as

$$SAGD(\omega) = \frac{\sum_{n=1}^{N} \sum_{m=1}^{M} \overline{GD}(\omega, \Delta\omega, \theta_n, \phi_m)}{NM}, \tag{2.19}$$

where $\overline{GD}$ is the group delay normalised by the maximum value. This is obtained under the same conditions as the $SATF(\omega)$, in order for it to be representative (Valderas *et al.*, 2006b).

2.2.3.2. *Correlation-based averages: Angular range*

The study, given by the fidelity factor, of the correlation between the input pulse to the transmitting antenna and the output pulse from the receiving antenna, is undoubtedly important. Another very useful analysis is the examining of the correlation between a pulse transmitted in a spatial direction, called the reference direction, and the pulses transmitted within a solid angle of interest Ω. For a given UWB antenna, the greater the value of Ω, the more difficult it will be to keep the correlation of all the signals above a threshold value. Once this has been defined, which will be called the Pattern Stability Factor (*PSF*), it will be possible to obtain

---

[b]In the article referred to, the maximum value is unique for all the directions of interest, therefore, the average is taken first, and it is then normalised, not the reverse: it is not normalised by the maximum for each direction.

an angular value, called the **angular range**, within which stable transmission is guaranteed. Angles smaller than the angular range will fall within the limits, while angles greater than the angular range will not. However, some other concepts must first be defined.

The **spatial correlation factor in the time domain** f^2 expresses the degree of correlation between the field radiated in one direction (θ_r, ϕ_r), and that radiated in a reference direction (θ_o, ϕ_o) (Dissanayake and Esselle, 2006). It can also be seen as the ratio of the correlated energy U_C between the chosen direction and the reference direction, to the total U_E. Its value is obtained using Eq. (2.20):

$$f^2(\theta_r, \phi_r; \theta_o, \phi_o) = \frac{U_C}{U_E} = \frac{\left[\int_{-\infty}^{\infty} e(\theta_r, \phi_r, t) \cdot e(\theta_o, \phi_o, t)dt\right]^2}{\int_{-\infty}^{\infty} |e(\theta_r, \phi_r, t)|^2 dt \cdot \int_{-\infty}^{\infty} |e(\theta_o, \phi_o, t)|^2 dt}, \quad (2.20)$$

where $e(\theta_r, \phi_r, t)$ and $e(\theta_o, \phi_o, t)$ represent the components radiated in the time domain in a given direction (θ_r, ϕ_r) and in the reference direction (θ_o, ϕ_o), respectively. The factor f^2 is less than or equal to 1, and is better the closer it is to 1.

By analogy, the **spatial correlation factor in the frequency domain** F^2 is calculated as the equivalent to f^2 in this domain.

$$F^2(\theta_r, \phi_r; \theta_o, \phi_o) = \frac{\left[\int_{-\infty}^{\infty} E(\theta_r, \phi_r, f) \cdot E^*(\theta_o, \phi_o, f)df\right]^2}{\int_{-\infty}^{\infty} |E(\theta_r, \phi_r, f)|^2 df \cdot \int_{-\infty}^{\infty} |E(\theta_o, \phi_o, f)|^2 df}. \quad (2.21)$$

The **frequency domain correlation pattern** $C(\theta_o, \phi_o)$ is the average value of the factor F^2 for the whole range of possible operating directions for the antenna within the solid angle Ω, with respect to the reference direction (θ_o, ϕ_o) (Dissanayake and Esselle, 2006). This is calculated as:

$$C(\theta_o, \phi_o) = \frac{\int_{\Omega} F^2(\theta_r, \phi_r; \theta_o, \phi_o)d\Omega}{\int_{\Omega} d\Omega}. \quad (2.22)$$

The **pattern stability factor** (*PSF*) is defined based on the frequency domain correlation pattern, and in turn is an average of $C(\theta_o, \phi_o)$ over all the possible directions that can be taken as reference directions within Ω (Dissanayake and Esselle, 2006). Its value is given by Eq. (2.23).

$$PSF = \frac{\int_{\Omega} C(\theta_o, \phi_o)d\Omega}{\int_{\Omega} d\Omega}. \quad (2.23)$$

It can be stated that $C(\theta_o, \phi_o)$ is calculated by integrating over the directions of interest, and the *PSF* is calculated by integrating over the reference directions.

An ideal UWB antenna would have the f^2, F^2, C and *PSF* factors equal to 1. In practice, a UWB antenna has good performance if its *PSF* is greater than 0.95.

It would therefore be very useful to translate these average correlations into UWB transmission angles that are reliable in terms of stability. For an arbitrary solid angle Ω, an antenna *PSF* drops as the antenna's bandwidth increases. Additionally, for a fixed bandwidth, the PSF also worsens as the value of Ω is increased. Given these limitations, the *angular range* $A(\theta_o, \phi_o)$ can be defined as the maximum angle of coverage within which the PSF is demonstrated to be > 0.95, within a radiation plane with respect to a reference direction (θ_o, ϕ_o), and for a specific bandwidth (Valderas *et al.*, 2008). Unless otherwise specified, the bandwidth is assumed to be such that the condition VSWR < 2 is fulfilled.

2.3. Simulation in the Time Domain

Electromagnetic simulation is very useful for analysing the figures of merit for a UWB antenna through an approximation, prior to making measurements. Time domain computation packages, such as CST Microwave Studio Suite provide a suitable platform for emulating the performance of pulsed UWB signals passing through a radiating element or within a complete transmission-reception system.[c]

This analysis allows the user to establish a specific antenna input signal $x(t)$ and then directly calculate the antenna transfer function, its group delay, etc. Since it works directly in the time domain, post-processing can be used to obtain the parameters described above, such as the fidelity factor and the angular range. Additionally, since the computing time drops as the simulation bandwidth increases for good mesh conditions, this solution is especially suitable for UWB antennas. This tool will therefore be used in the following chapters, in which UWB antenna design is covered.

---

[c]http://www.cst.com.

Chapter 3

Classification of UWB Antennas

David Puente * *and Daniel Valderas* †
Universidad Politécnica de Madrid (UPM)
† *CEIT and Tecnun, University of Navarra*

This chapter will review the various families of antennas used for this technology. They are classified into five major groups: helical, frequency-independent, log-periodic, horns and those derived from resonant antennas. The first four groups are based on classic wideband antenna designs which, by extension, have been used for UWB. Resonant antennas are of more recent origin and have awakened considerable interest, since they have characteristics which make them suitable for UWB use given the modifications that are often required. Most of these antenna types can be designed as either 3D or planar. Planar versions are easier to integrate with the other components of the communications system.

The aim of this presentation is to give the reader an overview of the different families of UWB antennas, and enable them to extrapolate the classifications described here to other designs.

3.1. Helical Antennas

The helical antenna was invented by Kraus in 1946 and was originally used for space communications (Kraus, 1950). It is constructed by winding a conducting wire around a cylinder of constant diameter. The feed point is located between one of the ends of the helix and a ground plane. The design parameters are therefore: diameter D, the separation between turns of the helix S, the diameter of the wire d, the number of turns N and the direction of winding (Fig. 3.1). It is a non-resonant, electrically large antenna.

Depending on the values of the design parameters, the helical antenna may work in one of two modes: normal or axial. The radiated field has two orthogonal components in phase quadrature, so circular polarisation may be obtained. For the

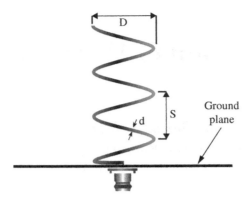

Fig. 3.1.   Geometrical parameters of the helical antenna.

helical antenna to operate in normal mode, the total length of the wire must be much less than the wavelength. In this mode, the radiation pattern has a maximum in the broadside direction. The helical antenna's radiation efficiency is low, because the wire's ohmic resistance is comparable to its radiation resistance. In normal mode, the bandwidth is not considerable, being typically 5%. Noguchi proposes using a helical antenna constructed with two parallel wires (Noguchi *et al.*, 2007) that are short-circuited at one end (folded structure). This raises the bandwidth by up to 12% at a working frequency of 760 MHz.

The axial mode resolves many of the deficiencies shown by the normal mode. For the helical antenna to work in this mode, its length must be comparable to the wavelength. The radiation pattern has a maximum in the end-fire direction. The values for radiation resistance normally lie between 100 and 200 Ω. A typical bandwidth of an octave can be obtained, for which the radiation resistance and the axial ratio between the components of the radiated field remain practically constant. The design criterion for optimal operation is that the following condition is fulfilled (Stutzman, 1998).

$$\frac{3}{4} < \frac{\pi D}{\lambda} < \frac{4}{3}. \tag{3.1}$$

There are variations on the helical antenna that improve on its broadband performance characteristics. Alsawaha *et al.* (2009) present a helical antenna for UWB that is wound around a sphere, instead of a cylinder as in the classic design. This gives a compact antenna with similar performance in terms of the main lobe beamwidth and directivity, but with greater bandwidth. It thus presents a VSWR < 2 between 2.9 and 5 GHz (bandwidth of 50%), and an axial ratio below 3 dB between 2.95 and 3.7 GHz (bandwidth of 24%).

3.2. Frequency-independent Antennas

Frequency-independent antennas are based on Rumsey's Principle, according to which, an antenna whose geometry is solely defined by angles is frequency-independent: the geometry is maintained at different scales which are determined by the operating wavelength. If this principle were strictly applied, it would require an antenna of infinite size for it to be truly frequency-independent. In practice, the geometry must be truncated for this to be achievable.

A concept related to the Rumsey Principle is that of the self-scaling antenna, which implies that the geometry is invariant when multiplied by a scaling factor K.

3.2.1. *Spiral antennas*

The simplest version of a balanced spiral antenna (see Fig. 3.2) is made up of two flat spirals in opposite directions. The geometry of a planar equiangular spiral is defined in polar coordinates as

$$r = Ae^{a\phi}, \tag{3.2}$$

which is a self-scaling curve for any scaling factor.

The arms are truncated at a length similar to the wavelength of the lowest frequency in the band. Designs can be obtained which give bandwidths of one decade. The feed point is located between the two closest ends of the spiral arms. The radiation pattern is of the $\cos(\theta)$ type, having poor directivity. Some applications require a unidirectional pattern. This can be achieved by mounting the arms of the spiral on a cone (Dyson, 1959).

The principal disadvantage of this type of antenna is the feed, which must be balanced over the whole of the band. Typically, a coaxial cable is used, which is an unbalanced line. It is therefore necessary to use a wideband balun. Furthermore, polarisation varies with frequency. At low frequencies it is linear, but

Fig. 3.2.   Spiral antenna.

as the frequency increases, elliptical polarisation is obtained, and it may become circular. The current distribution varies with frequency in such a way that at low frequencies the currents have longer paths which in turn generates dispersion. Spiral antennas are quite widely used in UWB applications. The performance of a circularly polarised spiral antenna having a microstrip-line balun is analysed by Mao *et al.* (2009). VSWR < 2 is obtained between 3.75 and 18.6 GHz. The axial ratio remains below 3 dB between 3 and 14.5 GHz.

3.2.2. Biconical antennas

3.2.2.1. 3D biconical antennas

The biconical antenna was invented by Lodge in the 1890s and extensively studied by Schelkunoff in the 1930s. It is made up of two opposing metal cones. The feed is located between the tips of the cones. Its use in UWB is based on the fact that it would theoretically be capable of providing frequency-independent impedance if it was of infinite length. The impedance is given by the following expression:

$$Z_{in} = Z_0 = 120 \ln \left(\cot \frac{\theta_k}{2} \right), \qquad (3.3)$$

where θ_k is the angle of the cone, as shown in Fig. 3.3, and Z_o is the characteristic impedance of the antenna which in this case is equal to the input impedance.

   In practice, the size is truncated which introduces reflections and limits the operating bandwidth. For example, a bandwidth of 6:1 can be obtained (McNamara *et al.*, 1984). This condition may be improved through graduated truncation. As for the spiral antenna, the phase centre remains constant.

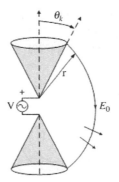

Fig. 3.3.   Geometry of the biconical antenna.

A UWB antenna inspired by the biconical antenna is proposed in Amert and Whites (2009). This is smaller than the conventional design. Miniaturisation is achieved through the use of a dielectric and pins. The pins short-circuit the two cones, in order to load the antenna with additional capacitance. This antenna has good matching characteristics, with VSWR < 2 over more than 7 GHz and good gain stability in the broadside direction.

The discone antenna is a variant of the biconical antenna (see Fig. 3.4). This is made up of a cone opposite a metal disk. The feed is supplied by a coaxial cable at the tip of the cone, which passes through the metal disk. The disk typically has a radius 0.7 times $\lambda/4$ at the lowest working frequency, the cone an angle of 25° (Stutzman, 1998). This antenna gives a stable radiation pattern over an octave and good matching over several octaves. The polarisation is linear and the radiation pattern is similar to that for a dipole.

A possible UWB application of the discone antenna, as shown in Fig. 3.4, is in constructing a radiating structure which can be reconfigured, as shown by Okamoto and Hirose (2008). In this, PIN diodes are used at the feed point to give the antenna different radiation patterns. Additionally, in this case, good matching characteristics are maintained, with a relative bandwidth greater than 76% for all the diode configurations.

A model which is a hybrid of the biconical antenna and the discone antenna was proposed for UWB use by Kim *et al.* (2005). Instead of a flat disk, a wide-angle ($\theta_k = 70°$) metal cone is used. This gives excellent broadband matching of 100:1 and acceptable omnidirectionality in the radiation pattern over the whole of the band.

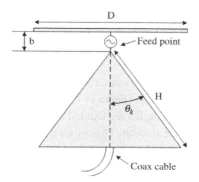

Fig. 3.4.   Discone antenna parameters.

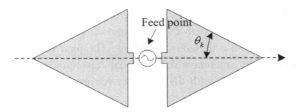

Fig. 3.5.   Diamond antenna.

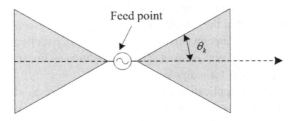

Fig. 3.6.   Bow-tie antenna.

3.2.2.2. *2D biconical antennas*

This family of antennas is based on the biconical antenna described above. One example inspired by this, although not frequency-independent, is the diamond antenna which is very popular in UWB systems. This antenna is made up of two opposing triangular plates. The feed point is located between them, as shown in Fig. 3.5.

The characteristics of the diamond antenna for UWB use are analysed in Lu *et al.* (2004). A variant of the design shown in Fig. 3.5 with rounded tips is also studied. The classic design provides a bandwidth of 2.8 GHz (2.7–5.5 GHz), while the variant's bandwidth is 3.7 GHz (2.7–6.4 GHz).

The bow tie antenna is another example of a 2D biconical antenna. This is produced by cutting through the biconical antenna transversally. It is therefore made up of two metal triangles opposite each other. The feed is located between the vertices. A bow tie antenna, as shown in Fig. 3.6, is shown in Kiminami *et al.* (2004), in which the metal triangles are located on different faces of the substrate, in order to obtain greater bandwidth. This gives a reflection coefficient less than −10 dB between 3.1 and 10.6 GHz.

3.3. Log-periodic Antennas

The geometry of a log-periodic antenna is self-similar, based on a basic cell which is repeated so that each cell is greater than the previous one by a scaling factor.

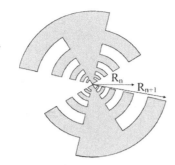

Fig. 3.7. Log-periodic antenna model invented by DuHamel.

This antenna has characteristics that repeat periodically as a logarithmic function of the frequency (giving it its name). The first ever model of an antenna of this type (DuHamel and Isbell, 1957) is shown in Fig. 3.7. For this to be self-scaling, the following condition for the ratio of two consecutive radii must be fulfilled:

$$\rho = \frac{R_{n+1}}{R_n},$$ (3.4)

where ρ is a constant.

The possible scaling factors are thus of the $K = \rho^m$, $m = \pm1, \pm2\ldots$ type. The antennas therefore present the same properties at frequencies f_0 and f_m with

$$f_m = \rho^m f_0.$$ (3.5)

If logs are taken of the two sides of the previous expression

$$\log f_m = m \log \rho + \log f_0$$ (3.6)

the initial proposition is proved.

The radiation pattern has a maximum in the direction perpendicular to the plane of the antenna, with a single lobe, which is linearly-polarised in the direction of the teeth. The lower limit of the working band depends on the length of the longest tooth.

Some modifications may be made to this design, in order to make it easier to construct. For example, the current distribution is concentrated at the edge of the antenna, so it can be implemented using just one conducting wire.

Another type of log-periodic antenna attributed to Isbell uses dipoles set out in a series of increasing size, as shown in Fig. 3.8. All the dimensions of the set are scaled by a factor, including the separations between the elements. In practice, scaling the diameters of the dipoles requires a considerable effort, so these remain constant.

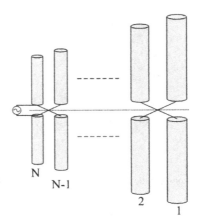

Fig. 3.8.   Log-periodic antenna with N dipoles.

The antenna is fed at the smallest dipole. For optimal operation, each dipole must be fed at a phase difference of 180° compared to its neighbours. At an arbitrary frequency, there is ideally only one dipole resonating. The dipoles which are larger than the resonating dipole act as reflectors, and the smaller ones act as directors, as in a Yagi antenna. The zone of the antenna which is resonating at a given frequency is called the active region. This zone moves along the log-periodic antenna in the working band. The phase centre therefore changes with frequency, which causes the dispersion that makes this type of antenna less than optimal for UWB. The bandwidth is set by the longest and shortest dipoles. This type of log-periodic antenna has high directivity, e.g., 10 dB. It is mainly used for radio links in the HF band.

The UWB characteristics of a dipole-based log-periodic antenna with an average gain of 8 dB_i are evaluated by Merli *et al.* (2009). It displays good matching, with VSWR < 2 between 4.2 and 10.6 GHz. To reduce the dispersion, it is proposed that a linear ratio is used between the frequencies in Eq. (3.6) instead of a logarithmic ratio. As a consequence, the condition set out in Eq. (3.4) is no longer fulfilled, and the antenna is not strictly self-scaling. There is, therefore, a reduction in impedance bandwidth, although this may be corrected by using more dipoles when constructing the antenna.

3.4. Horn Antennas

3.4.1. *3D horn antennas*

3D horn antennas were invented by J. C. Bose in 1890. They are constructed as a widening of the waveguide that feeds the antenna. The cut-off frequency of the

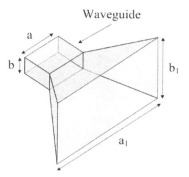

Fig. 3.9. Pyramid horn antenna.

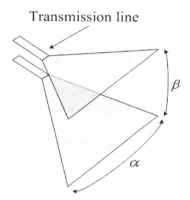

Fig. 3.10. Conical horn antenna.

fundamental mode may be controlled by changing the dimensions of the guide. The operating principle is based on radiating apertures. A typical design is a pyramid horn, illustrated in Fig. 3.9. This provides a high gain, between 5 and 15 dB$_i$, so it is especially used for point-to-point links, and also in arrays.

The main disadvantage of these antennas is the fact that the gain is not stable over frequency. They also tend to be electrically large, of considerable size, typically around one wavelength of the lower operating frequency. Other critical aspects are the reflections produced at the end, and the diffraction at the edges. A typical technique for reducing this effect is to have rounded edges, which attenuates the side lobes and gives a radiation pattern that is more stable over frequency.

Horn antennas are very popular for UWB. Although some authors do not consider their phase centre to vary with frequency (van Cappellen *et al.*, 2000), a certain amount of phase centre drift (depending on the magnitude of the pulse and the bandwidth) and the consequent ringing cannot be avoided (Chen, 2005).

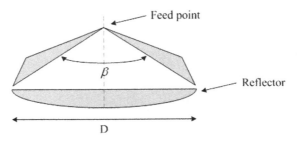

Fig. 3.11.   Parameters of an IRA.

The conical horn antenna is another widely used design, consisting of two triangular sheets (see Fig. 3.10). A TEM or quasi-TEM mode is previously excited over a two-conductor transmission line. Its geometry is characterised by the angle α of the triangular plates and the angle β between them. Different input impedances and directivities may be obtained, depending on the values of these two angles (Lee *et al.*, 2004).

To achieve greater levels of directivity, the horn antenna is placed as the feed for a parabolic reflector to form an Impulse Radiating Antenna (IRA), as shown in Fig. 3.11. This antenna gives a practically constant gain of up to 25 dB$_i$ (Baum and Farr, 1993), and this can be modified by moving the feed.

3.4.2. *2D horn antennas*

This type can be considered to be a specific case of the 3D antennas, with angle $\alpha = 0$. The best known one in this group is the Vivaldi antenna, as shown in Fig. 3.12, which was introduced by Gibson in 1979. This can theoretically provide infinite bandwidth, although in practice this is limited by its size, manufacturing techniques and the type of feed. The feed is provided through a microstrip-line and a slot-line. The transition between the two lines causes a reduction in bandwidth. The most common solution is to place a wideband balun between them.

This radiator presents good phase linearity and a low level of distortion in the working band. The radiation pattern has quite good directivity, in the region of 10–15 dB$_i$. The antenna's shape and dimensions determine the side lobe level. They also serve to control the different resonant frequencies that define the antenna's broadband response. The level of the co-polar component and the beamwidth can be controlled by changing the radius of the curved profile of the metallic parts. So, for example, Greenberg *et al.* (2003), studied a Vivaldi antenna with optimised interior and exterior profiles. This achieved lower cross-polarisation and better

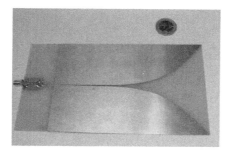

Fig. 3.12.   Vivaldi antenna constructed on an FR-4 substrate.

levels of directivity in the E- and H-planes at high frequencies in the working band, in comparison with the classic Vivaldi antenna.

A Vivaldi antenna for UWB, designed to cover the whole of the UWB band assigned by the USA's FCC (3.1–10.6 GHz), was analysed by Hood *et al.* (2008). However, the gain varied considerably, between 0 dB$_i$ at the lower edge of the band and 6 dB$_i$ at the upper edge. This weakness was fundamentally due to the fact that a balanced feed was not used. To achieve this, an antipodal design was employed, which used both faces of the dielectric.

3.5. UWB Antennas Derived from Resonant Antennas

The previous sections have introduced versions of UWB antennas that were derived from non-resonant antenna families. They were, for example, derivations of antennas considered to be frequency-independent, such as the bow-tie, which is derived from the biconical antenna, and the Vivaldi, which is derived from horn antennas.

This section starts with a typical narrowband resonant antenna, the linear monopole, which is modified to give UWB monopoles. The slot-antenna, which is converted into a UWB antenna by changing its shape, is also discussed. Although the final distinction between a truncated frequency-independent antenna and, for example, a UWB monopole, may not be clear, the authors believe that this way of presenting the different types of antennas will be enlightening.

3.5.1. *3D monopoles*

While maintaining their topology in terms of volume, the 3D monopoles consist of a radiating element (typically a metal sheet) over a perpendicular ground plane.

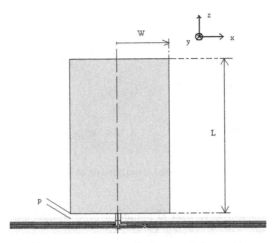

Fig. 3.13.   Planar monopole antenna over perpendicular ground plane and its geometrical parameters.

The simplest is the linear monopole, made up of a radiating element of length $\lambda/4$, perpendicular to the ground plane.

This is a resonant antenna which is not suitable for UWB. Dubost and Zisler (Dubost and Zisler, 1976, 128–129) therefore studied techniques to increase the bandwidth by widening the antenna. This provided a satisfactory response to requirements for UWB and broadband applications in general. This type of 3D monopole is called a Planar Monopole Antenna (PMA) over perpendicular ground plane, or simply a planar monopole, for short[a], as shown in Fig. 3.13.

Like all monopoles, the 3D monopoles exhibit good phase linearity (Kerhoff *et al.*, 2001), very high efficiency (Schantz and Fullerton, 2001) and very high bandwidths, for example 14:1 (Qiu *et al.*, 2005), can be achieved. Additionally, techniques may be used to filter out certain frequency bands. Their radiation pattern is relatively omnidirectional. One should also emphasise that the cost of manufacturing this type of antenna is low.

The lower limit of the operating band for a monopole is defined by the longest electrical path that the currents generated on the surface of the monopole can travel. The upper limit depends on other factors, such as the shape of the monopole and the separation from the ground plane. Various techniques for increasing the differences between these limits are discussed below.

---

[a]This should not be confused with the full 2D monopole antennas or PCB monopole antennas, for which the ground plane is parallel to the monopole.

3.5.1.1. *Modifications to the geometry*

3.5.1.1.1. Euclidean shapes

One technique for increasing the bandwidth so that it conforms to UWB specifications consists of testing different Euclidean shapes. This originated in Brown's work (1952), which was the first study of triangular monopoles. The performance of a triangular monopole is limited, with a typical bandwidth of 20%, about 1 GHz (Kerkhoff *et al.*, 2004). The principal advantage is that it can be connected directly to the feed without short-circuiting to the ground plane as a widening at the base of the monopole is avoided.

However, the most representative antenna of the planar monopoles with Euclidean shapes is the square antenna (Agrawall *et al.*, 1998). Its bandwidth is around 75%, with VSWR < 2 in the S-band.

A variant of this is the bow-tie shaped monopole (see Fig. 3.14), which has a bandwidth similar to the square antenna, but the lowest frequency in the band is reduced, as the electrical path of the currents circulating around the outer edge is increased (Chen, 2000a).

Trapezoidal monopoles (see Fig. 3.15), for which the ratio of the lengths of the top and bottom sides is greater than 1, achieve some improvement in bandwidth (80%) for the same reason as the bow-tie monopoles (Chen *et al.*, 2000b).

Of all the Euclidean shapes, the circular and elliptical ones give the highest bandwidths. So, for example, an optimised elliptical monopole for which the ratio

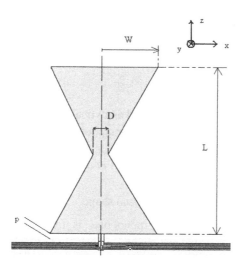

Fig. 3.14. Bow-tie-shaped monopole.

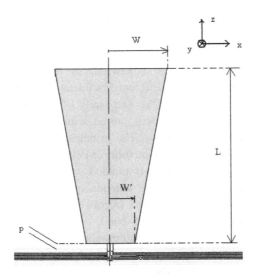

Fig. 3.15.   Trapezoidal monopole.

of the longer axis to the shorter one is 1.1, provides a bandwidth of 10.7:1 (Agrawall *et al.*, 1998). Another example is the commercial circular antenna designed by Honda (Honda *et al.*, 1992), with a bandwidth of 8:1 (90–770 MHz).

Some authors consider that the radiation pattern of elliptical antennas, as shown in Fig. 3.16 deteriorates more than in rectangular monopoles (Ammann, 2000). Although the conditions of the comparison were not completely equivalent, it does seem that there is a compromise between radiation pattern stability and impedance bandwidth.

3.5.1.1.2. Computer optimisation

Given the lack of experimental data and analytical models, another way to tackle design problems is through the use of computer optimisation techniques. Typically, this is carried out by applying Genetic Algorithms (GA), which allow more elaborate shapes to be developed (Kerkhoff *et al.*, 2004).

3.5.1.1.3. Partial variation on a Euclidean shape

Another way of studying planar monopoles is by making modifications to geometries that have already been studied. This is the case when the profile of the lower edge is modified to increase bandwidth, for example by making a bevelled notch or cut (Ammann, 2001), as shown in Fig. 3.17. It should be emphasised that

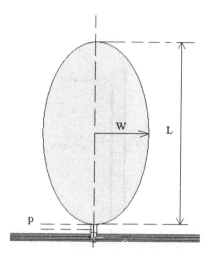

Fig. 3.16. Elliptical monopole.

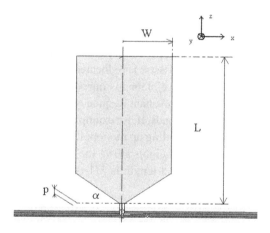

Fig. 3.17. Monopole with bevel-shaped lower edges.

the characteristics and performance of planar monopoles are determined, above all, by the configuration of the antenna close to the feed point.

3.5.1.2. *Changes in current distribution*

In this case, various techniques are used in order to alter the current distribution, as shown below.

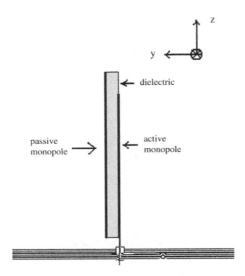

Fig. 3.18.   PMA with coupled parasite rejecting the unwanted band (side view).

3.5.1.2.1. Use of parasitic elements

The use of parasitic elements is one possibility. This means that elements without a wired connection to a feed (i.e., passive) are located close to elements connected to the feed (i.e., active). The passive elements introduce mutual coupling effects, although they must have a similar resonant frequency to the active ones in order to achieve a partial overlap between bands. If, for example, the antenna is implemented on a PCB, one method consists of piling up monopoles in parallel layers. However, when the monopoles have similar heights above the ground plane (Chen, 2000b), the frequency for which the vertical length is $\lambda/4$ is filtered out of the band (see Fig. 3.18).

Other solutions based on these techniques focus on the case in which the active monopole is narrow and the parasite has a square or rectangular Euclidean shape of known characteristics. In this way, bandwidths of 8:1 have been obtained (Chen and Chia, 2000b). However, one of the monopoles must be considerably longer than the other in order to prevent any frequencies being filtered out.

3.5.1.2.2. Use of a short-circuit pin

Another design technique for planar monopoles consists in using short-circuit pins in the areas with high current density (see Fig. 3.19). In addition to providing greater bandwidth, this is also useful for making the antenna more compact. Bandwidth

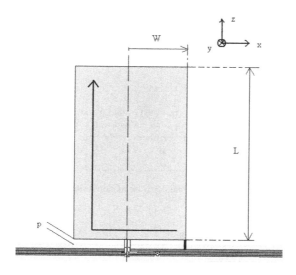

Fig. 3.19.   Planar monopole with short-circuit and another mode excited.

is increased from 75% in the S-band, up to 110% (Ammann, 2000). If this is implemented on a dielectric, the size is reduced by up to 50% (Lee *et al.*, 1999).

3.5.1.2.3. Asymmetric feed

An asymmetric feed may also be used. This allows the matching to be changed, and improves bandwidth at the upper end. It consists in moving the feed point along the horizontal axis, so that it is not on the axis of symmetry. Unlike short-circuiting the antenna, this method does not cause additional modes to be excited at lower frequencies, and so does not result in a more compact antenna. However, it is similar in that the current symmetry is lost. The radiation pattern therefore loses its omnidirectionality, since an offset is introduced in the horizontal plane, as in the short-circuit case (Ammann and Chen, 2004).

3.5.1.2.4. Double feed

This technique is based on using two feed points, instead of the more usual one. The horizontal currents are reduced. In this manner, the horizontal currents are reduced (Cabedo-Fabres *et al.*, 2003). A square planar monopole is proposed by (Daviu *et al.*, 2003) with two feed points which are symmetrically placed with respect to the centre of the antenna. Cross-polarisation is reduced because this system reinforces the excitation of vertical modes.

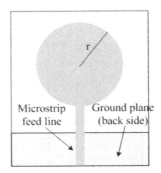

Fig. 3.20. Geometry of the printed circular monopole with a microstrip feed.

3.5.2. *2D resonant antennas*

3.5.2.1. *Full 2D monopoles*

Full 2D monopoles consist of a metal patch on one face of a dielectric board, with the ground plane parallel to it, usually on the other face of the printed circuit. The feed is normally provided through a microstrip line having the same ground plane as the monopole. When both parts (monopole and ground plane) are coplanar on the same surface, the feed is normally provided by a coplanar waveguide.

In the past, implementing antennas of this type was usually conditional on the microstrip line's limited bandwidth, which was not more than 50% of the radiating element bandwidth. This limitation has now been overcome and the line's bandwidth may reach 90% of that of the patch (Garg *et al.*, 2000).

One example is the antenna shown in Liang *et al.* (2005), which consists of a circular monopole printed on a substrate, as illustrated in the Fig. 3.20.

The radius is optimised in order to maximise bandwidth and achieve full cover of the FCC's UWB band. The radiation pattern remains reasonably omnidirectional throughout the whole of the band.

As mentioned, a different type of line may be used to provide the feed to the patch. Liang *et al.* (2005) studied a circular patch antenna with a coplanar waveguide feed, as shown in Fig. 3.21. This gives a bandwidth at −10 dB from 3.27 to 12 GHz. The dimensions of the patch can be used to control the first resonant frequency, and therefore the lower end of the band. The radiation pattern remains acceptably omnidirectional.

3.5.2.2. *Slot antennas*

Just as the full 2D UWB monopole antennas are a modification of a resonant monopole, so the UWB slot antennas described below are a variation on the slot

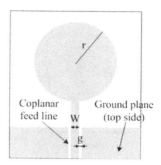

Fig. 3.21.   Geometry of the printed circular monopole with a coplanar waveguide feed.

antennas that have traditionally been considered to be narrowband. They can be seen as a complementary version of the full 2D monopoles. They consist of an aperture in the metallised face of a substrate, and are normally fed through a coplanar waveguide or a slot-line. The shape of the aperture or slot may be rectangular, triangular, elliptical, etc. Circular apertures result in the resonant modes being closer in frequency, which facilitates greater overlapping between these modes and consequently a widening of the bandwidth (Kumar and Ray, 2003).

Slot antennas provide high bandwidths, although generally not sufficient to cover the whole of the UWB band defined by the FCC. In addition to broadening the slot, the feed is therefore usually loaded with elements that increase the bandwidth, such as patches or stubs. A very useful characteristic is that they confine the electrical field in the dielectric. This is necessary in devices that require the antenna to be closely integrated with the rest of the circuit.

Figure 3.22 shows the slot antenna studied in Sze and Shia (2008). The antenna's bandwidth may be controlled by changing the position of the patch located on the other face of the dielectric, to give VSWR < 2 between 3.8 and 10.7 GHz. Slots may also be made in the metal patch, in order to add notch-bands.

An ellipse-shaped slot antenna is shown in Li *et al.* (2006). Different types of feed are studied, using a microstrip line and a coplanar waveguide terminated by a U-shaped stub. The dimensions of the stub may be used to control the lower limit of the working band. In the 3–10 GHz range, the gain varies between 2 and 7 dB$_i$ and the return loss is below −10 dB.

3.6. Conclusions

Given the review of antenna types, one can ask which type of antenna is most appropriate for each specific UWB application. It has been demonstrated that each type presents advantages and disadvantages.

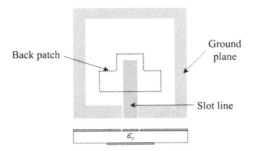

Fig. 3.22.   Geometry of a square slot with a coplanar waveguide feed.

To summarise, in frequency-independent antennas, there is a tendency for the current to become attenuated as the fields are radiated — these would be electrically large antennas. Thus, when the edge of the antenna is reached, resonance due to reflection is negligible, as only a progressive wave is set up. Fields at different frequencies are radiated by different parts of the antenna.

The major disadvantage this presents is that the phase centre is not well-defined, and/or the useful radiating area where the currents circulate "travels". This phenomenon is more acute in the case of log-periodic antennas. In a UWB system, all the frequencies radiate at the same time, so the lag introduced by a log-periodic antenna, and the consequent ringing, would be unacceptable (Soergel *et al.*, 2003). The antennas would be inherently dispersive and the necessary specifications would not be fulfilled (Yazdandoost and Kohno, 2004). This effect applies, though to a lesser extent, to broadband horn antennas (Chen, 2005). The use of these antennas, as for helical antennas, is focussed on eminently directional applications which restrict their scope.

The supposedly frequency-independent antennas present other disadvantages. In the case of spiral antennas, in which a change of scale is equivalent to turning the antenna, the radiation pattern turns in the same way with frequency, although it maintains its shape (Mayes, 1992). Additionally, the axial ratio of polarisation changes with frequency (*ibid.*). This often defines the lower limit of the band. Similarly, the feed must be balanced, which requires a wideband balun. This increases the number of restrictions and the complexity of the layout of the coaxial cable (Cardama *et al.*, 1998). At the same time, studies have been carried out that compare monopoles and spirals having a rear cavity which make it clear that the ringing introduced by spiral antennas is worse than for monopole antennas (Licul and Davis, 2003). Lastly, it should be emphasised that the solutions which imply frequency-independence, as for horn antennas, are too large to be compatible with the demands of portable UWB terminals (Chen, 2005).

As mentioned, the family of UWB monopoles display good phase linearity (Kerkhoff *et al.*, 2001) and high radiation efficiency (Schantz and Fullerton, 2001). When these antennas are appropriately designed to compensate for the fact that they are resonant, they offer exceptional bandwidths, e.g., 14:1 (Qiu J. *et al.*, 2005). These monopoles can have electrical lengths which are shorter than $\lambda/4$ at the lowest operating frequency (Agrawall *et al.*, 1998). When very high bandwidths are not required, significant profile reductions can be achieved (Kerkhoff and Ling, 2002), by making the antenna parallel to the ground plane and thus suitable for implementation in terminal devices (Wong *et al.*, 2004c). These designs can also filter out specific frequencies, in order to prevent interference with other existing applications. This interference is the main reason for some distrust of this technology. Filtering may be carried out directly through the design of the antenna itself (Su *et al.*, 2004). Additionally, this type of antenna presents radiation patterns that are relatively omnidirectional over the operating band (Kerkhoff and Ling, 2003). Lastly, the cost of implementing is low. Due to the fact that these monopole antennas are so technologically interesting, future chapters will tackle how they can be appropriately designed. All this will not only be applicable to UWB applications, but also to any transmission system requiring high bandwidth.

Chapter 4

UWB Monopole Antenna Analysis

Daniel Valderas and Juan I. Sancho
CEIT and Tecnun, University of Navarra

The previous chapter gave a general description of the different typologies of broadband antennas for UWB applications and listed the principal reasons why broadband monopole antennas offer a desirable solution in terms of both phase linearity and compactness. This chapter establishes a basis for studying these antennas. Based on current distributions in Planar Monopole Antennas (PMAs), an analogy is defined between this type of antenna and a transmission line with an arbitrary load at its end. Using this as a starting point, broadband matching techniques will be applied within the scope of transmission line theory. This will allow the development of a methodology for designing new monopoles whose bandwidth can be controlled through design.

4.1. Introduction

The classic theory of solving linear radiating elements involves the study of current distribution. These studies often assume a sinusoidal current distribution. One of the ways to increase bandwidth, which is an absolute requirement of UWB technology, is to widen the conductor. In order to understand how any changes in the geometry of the antenna will affect its behaviour, knowledge of the current distribution in this new situation is vitally important.

4.2. Current-conductive Parts on Planar Monopole Antennas

4.2.1. *Currents parallel and perpendicular to the ground plane: A working hypothesis*

The general appearance of current distribution for a square PMA having the feed in the centre of one edge and located over a perpendicular ground plane is shown

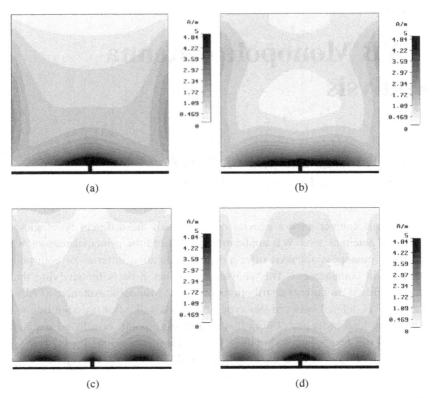

Fig. 4.1.   Maximum current density distribution (A/m) in a 45 mm × 45 mm PMA for (a) 1.5 GHz
(0.225 λ), (b) 4 GHz (0.6 λ), (c) 7 GHz (1.05 λ) and (d) 10 GHz (1.5 λ).

in Fig. 4.1. As can be seen, the current is concentrated in the bottom edge of the
monopole, as well as at the left and right-hand edges.

Thus, making a slot in the centre of a rectangular PMA (Evans and Ammann,
2003) does not significantly reduce the impedance bandwidth and improves other
factors such as weight, wind resistance and cost (Fig. 4.2).

Other authors (Chen *et al.*, 2002) propose annular antennas based on disks
(Fig. 4.3) that, to a considerable extent, maintain their impedance bandwidth with
a reduction in area of up to 64%. It can also be observed that the differences in the
radiation pattern are not significant, especially at low frequencies.

Likewise, for certain distances above the ground plane, when a square PMA
is made annular, with a hole in the centre, the impedance bandwidth remains
basically unchanged (Chen *et al.*, 2003). Comparing the different cases for a slotted

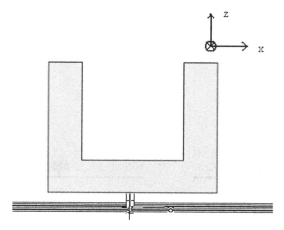

Fig. 4.2.   Square PMA with a central slot.

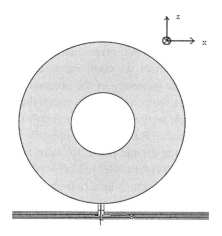

Fig. 4.3.   Circular PMA with hole in the centre.

square PMA, differences are observed between a U-shaped antenna and an annular antenna, where the first case is a more aggressive cut.

Furthermore, Fig. 4.4 shows the current amplitude of horizontal and vertical components, with respect to the ground plane, for a given representative frequency and PMA. This shows that the horizontal currents are located in the lower part, while the vertical currents are concentrated in the sides and around the feed.

A working hypothesis can therefore be established for current distribution:

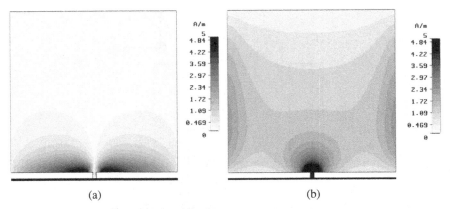

(a)   (b)

Fig. 4.4.   Maximum current density distribution (A/m) in a 45 mm × 45 mm PMA at 1.5 GHz (0.225 λ); (a) horizontal components and (b) vertical components.

In an antenna whose structure is monopolar over a perpendicular ground plane, in the edge closest to that plane, the current components parallel to the plane are greater than those perpendicular to it.

The phenomenon described is also representative of any PMA which is not square, in which the proportion of the components may be different from that shown here, depending on the geometry of the antenna and on the shape of the lower edge. Figure 4.5 shows the comparison between the currents parallel to the ground plane and those at right angles to it, for a monopole having a different shape. Although the current is forced to follow a more upwards path along the bevelled edge, in the edge closest to the ground plane, the currents parallel to the ground plane predominate over those at right angles to it. This phenomenon does not occur near the feed point, where the current is locally forced to follow a vertical path.

4.2.2. Non-radiating currents in a PMA

In a more rigorous study, apart from analysing which parts efficiently conduct current in a PMA, one should evaluate whether all the currents radiate to the same extent.

As is well known, any monopole is studied as derived from a dipole in which one half is suppressed by the presence of an infinite conducting plane. This can be simply explained in terms of the method of images. The fact that there is a ground plane allows one to go from the current distribution of a PMA to that of an equivalent Planar Dipole Antenna (PDA), as shown in Fig. 4.6.

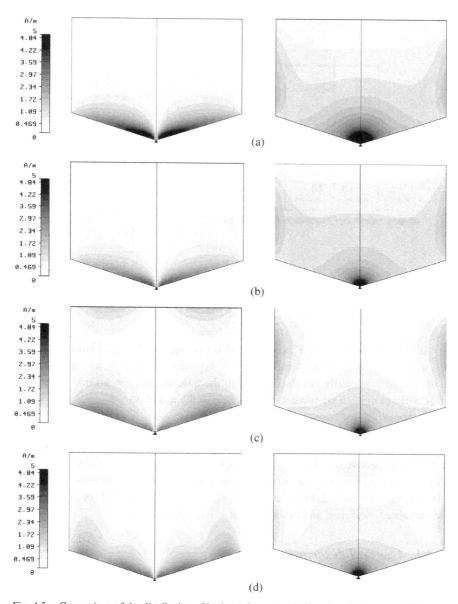

Fig. 4.5. Comparison of the distribution of horizontal currents (left) and vertical currents (right) in a PMA with a bevelled edge at (a) 1 GHz, (b) 2 GHz, (c) 5 GHz and (d) 10 GHz. Height = 45 mm, separation from the ground plane 1 mm, Width = 60 mm. Height of bevel at far edge is 10 mm.

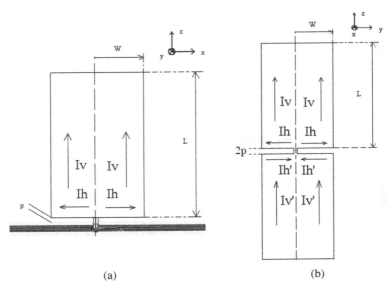

Fig. 4.6.　Passing from a PMA to its equivalent PDA with its current distributions both horizontal (I_h) and vertical (I_v). Image currents are primed, e.g., Iv'.

If the separation p from the ground plane is electrically small with respect to the wavelength under consideration, the fields produced by the horizontal currents will remain principally confined between the antenna and the ground plane and thus will not make a significant contribution to the radiation.

4.3. Transmission Line Model for UWB Monopole Antennas

Based on the above observations, a descriptive model may be defined for the behaviour of a conventional monopole antenna over a specified range of frequencies, based on transmission line theory. This allows monopole-type antennas to be explained, designed and optimised for UWB applications.

4.3.1. *General description*

Transmission line theory postulates that, while the separation between the two conductors making up the line is small enough in electrical terms, radiation could not be considered in calculations. However, in light of the argument in the previous section, we can conclude that, in the part close to the ground plane, due to the predominance of non-radiating currents parallel to the plane, the antenna may be

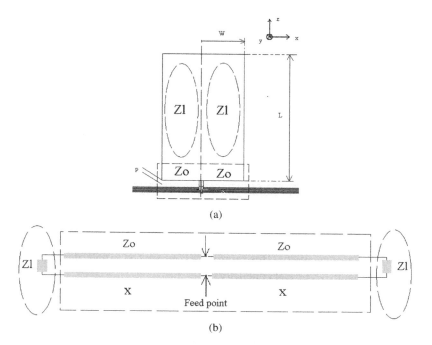

Fig. 4.7. Analogy between PMA and TLM.

Table 4.1. Correspondence between PMA and TLM.

PMA	TLM
Lower profile contour and separation	Zo
W	x
Radiating element	Zl

modelled by a transmission line. This "line" will be represented by its characteristic impedance Zo and its length x and will be loaded with an arbitrary impedance that represents the radiating part of the antenna. We will call this model the Transmission Line Model (TLM) applied to monopole antennas.

Figure 4.7 reflects the parallels between PMA and TLM. Table 4.1 shows the correspondence between their parameters. Although the analogy has been established for a rectangular PMA, the model is not limited to this case, since the results may be generalised.

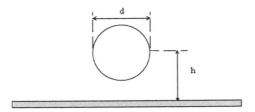

Fig. 4.8.   Cross-section view of a circular transmission line above ground plane.

4.3.2. *Description of the model*

4.3.2.1. *Transmission line*

The transmission line is made up of the lower part of the antenna and the ground plane. As an approximation, Fig. 4.8 shows a transmission line of circular cross-section, above the ground plane. Its characteristic impedance Zo is defined by Eq. (4.1) (Rudge *et al.*, 1986). This expression corresponds to half of an equivalent two-wire line with cylindrical conductors.

$$Zo = 60 \ln \frac{4h}{d}. \tag{4.1}$$

If the cross-section is not circular, an equivalent radius can be assigned to it. Its behaviour would be the same in qualitative terms.

In the PMA, the area occupied by the section that conducts current (which may serve to establish the equivalent radius) is not defined by its physical dimensions. It will therefore not be possible to reach a specific value for Zo. However, a model of this type allows the geometrical variables on which it depends to be observed. Thus, for example, when the height above the ground plane increases, Zo also increases. This variation is also logarithmic, i.e., equal increments in h will have greater effects at small values of h.

4.3.2.2. *Radiating structure*

If the transmission line is eliminated from the whole arrangement, the remaining part will be referred to in the model as the radiating structure. From the point of view of the model, it may be represented by an impedance Zl (Fig. 4.7(b)).

4.3.3. *Purpose of the analogy*

Given that the line in Fig. 4.7 is placed before the radiating structure itself, it could be considered to be part of the matching network, although its parameters are unknown. Seen in this way, the study of UWB monopole antennas and their optimisation would now translate into the application of broadband matching techniques using transmission lines loaded with arbitrary impedances. These techniques can be applied without the need to use known numerical values, but instead by analysing the trends of changes in the antenna impedance in the Smith Chart, which are caused by changes in the geometry of the transmission line.

4.3.4. *Graphical approach: The Smith Chart*

Problems that define the transmission line parameters necessary for matching a series load Zl are called transmission line synthesis problems. The techniques for tackling this type of problem have been extensively covered in previous studies (Collin, 1956; Klopfenstein, 1956).

Ordinarily, when talking about matching, one is working with a limited range of frequencies. In addition, when the term "broadband matching" is used, it is reserved for real loads and not for matching arbitrary impedances that vary with frequency. At most, an analytical limit is established for the bandwidth that can be obtained by matching a complex and arbitrary load (Fano, 1950). When more elaborate attempts have been made, the appropriate approach for achieving this type of matching is based on the Smith Chart. In fact, this tool has been used as the most suitable one for delimiting the zone in which it is possible to match an arbitrary impedance (Zentner *et al.*, 2003; Vendelin *et al.*, 1990).

In this context, a transmission line joined to a purely resistive load Zl, such that $Zo < Zl$,[a] displaces it towards the centre of the graph by rotating clockwise (Fig. 4.9). Depending on the value of Zl, values for Zo (that define the centre of rotation) and for length of line x (length of the arc over the chart) exist, which cause matching to the value of the impedance at the desired frequency (Zin, f_o). In the example, a $\lambda/4$ transformer is used. The analytical solution to the problem is given by Eq. (4.2).

$$Zin = Zo\frac{jZo \cdot \tan(\beta x) + Zl}{jZl \cdot \tan(\beta x) + Zo}. \tag{4.2}$$

If we start from a line of given dimensions and study how this line behaves in terms of frequency, we observe that the rotation produced is the same as that described above. As we have mentioned, the most clearly identifiable point on the chart is

---

[a]The reference impedance is assumed to be 50 Ω.

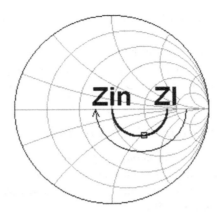

Fig. 4.9.   Matching to 50 Ω according to a resistive load *Zl* of 180 Ω at 4.8 GHz with a line having *Zo* of 95 Ω and length 15.6 cm.

given at the frequency at which the length of the line is $\lambda/4$:

$$Zin = Zo\frac{jZo\tan\left(\frac{2\pi}{\lambda}\frac{\lambda}{4}\right) + Zl}{jZl\cdot\tan\left(\frac{2\pi}{\lambda}\frac{\lambda}{4}\right) + Zo} = \frac{Z_0^2}{Zl}. \qquad (4.3)$$

At this point, the effect of the line is translated into a rotation of 180°, at the end of which, we expect the matching to have improved. At the frequency at which its length is $\lambda/2$, $Z_{in} = Z_l$ once again, which corresponds to a rotation through 360°. When comparing the effect of two different lines over the same load in order to select the one that improves matching, by comparing their effects on the Smith Chart, the decision is obvious.

If, instead of a purely resistive load *Zl* we have a complex load, this introduces variability with frequency for the load to be matched (Fig. 4.10(a)). If broadband matching is intended, instead of a single starting point, there will be a trace or set of points. When the transmission line is introduced, the rotations or turns mentioned above give rise to loops (Fig. 4.10(b)). Once again, points $\lambda/4$ (A and B) are the key points for defining the rotation and therefore the loop, which will be more or less closed and centred in the chart, depending on the initial situation and the transmission line that is applied. In the example, the grey trace is the one that gives the better broadband matching.

Thus, when confronting complex matching problems, whether due to the variability of *Zl* with frequency, or due to transmission lines having variable *Zo*, the analytical method definitively gives way to the graphical method, and the Smith Chart becomes an important visual tool.

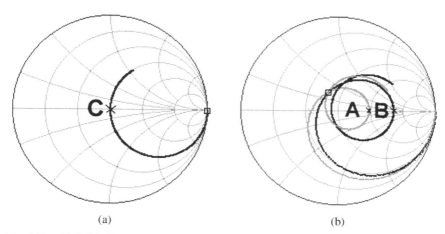

(a)   (b)

Fig. 4.10. (a) Smith Chart (0–11GHz) for a Zl made up of a series RLC circuit, where $R = 50\,\Omega$, $L = 1$ nH and $C = 1$ pF. (b) Smith Chart (0–11GHz) for the same Zl as in (a), preceded by a line of length 15.6 mm. Grey for $Zo = 70\,\Omega$, black for $Zo = 95\,\Omega$. Points A and B represent the frequency at which the line is $\lambda/4$ and correspond, in frequency terms, to point C.

4.4. Design Based on TLM

If the current distribution condition is respected, the model can be applied to several types of broadband monopoles to ensure that they should be suitable for UWB applications. A series of cases are given below.

4.4.1. *Design of an UWB-PMA antenna with a given bandwidth*

The principal advantage of PMAs is the simplicity of their construction. If the aim is to give the antenna a specific bandwidth, the technique of a multisection impedance transformer can be used in the zone defined as a transmission line (Pozar, 2005). This translates into a stepped profile for the edge closest to the ground plane. The lowest operating frequency is given by the dimensions of the PMA's outline shape. A detailed study of this case is given in Chap. 5.

4.4.2. *Design of an UWB-PMA antenna having a maximised bandwidth*

To achieve the greatest possible bandwidth, the Zo of the transmission line should be continuously variable between the impedance of the feed port and Zl. This variation should be as smooth as possible. This translates into using profiles whose height over the ground plane changes continuously. These are implemented by the edge closest to the ground plane having a rounded or bevelled profile. A detailed study of this case is given in Chap. 6.

4.4.3. *Design of omnidirectional UWB antennas*

A possibly important aspect of UWB antennas is the omnidirectional behaviour of the radiation pattern in the horizontal plane with variations in frequency. In the cases mentioned above, symmetry and therefore omnidirectionality are lost. In order to achieve this, folded monopole antennas (FMA) or revolution monopole antennas (RMA) are used based on a PMA. The change in the shape of the PMA antenna causes Zl to change, but the broadband matching techniques are still valid. Chapters 7 and 8 study how these should be designed.

4.4.4. *Design of directional UWB antennas*

To design a monopole with high directionality at specific frequencies, the edges would be brought together, so as to cause an acute asymmetry that favours one direction. An example of this type of antenna will be seen in Chap. 7.

4.4.5. *Design of 2D PCB antennas for UWB*

If the design is conditioned by the need for full 2D antennas for integration in UWB terminals, geometrical requirements leave little freedom for configuring the stability of the radiation pattern. In exchange for this, very low profile antennas are achieved. Chapter 9 will offer guidelines for designing them, based on TLM. Figure 4.11 illustrates the reference for the subsequent chapters.

In short, the TLM model remains valid as long as the new configurations do not affect the working hypothesis that establishes the current distribution. The impedance bandwidths obtained are analogous, whether the antenna is folded, revolving or simply planar. As a consequence, the radiation pattern conditions define the starting geometry (Zl) and applying TLM allows the required bandwidth for the specific UWB application to be obtained (Zo and x by design of the profile of the lower edge). Two introductory examples are presented in the next section.

4.4.6. *Case study 1: Semi-rectangular planar monopole case*

Figure 4.12(a) shows a PMA for which the impedance plot in the Smith Chart, as a function of the parameters p and W, can be studied in the light of the TLM model. To simplify the explanation, half of a conventional rectangular PMA is used.

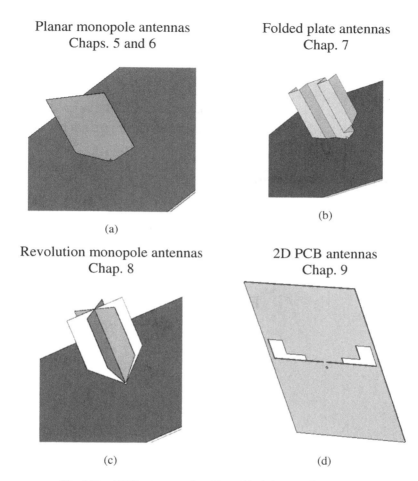

Planar monopole antennas
Chaps. 5 and 6

(a)

Folded plate antennas
Chap. 7

(b)

Revolution monopole antennas
Chap. 8

(c)

2D PCB antennas
Chap. 9

(d)

Fig. 4.11. UWB antennas to be addressed in their respective chapters.

4.4.6.1. *Parametrical transformation according to TLM*

The initial values Zl and Zo are approximate, since the important aspect is not the impedance plots themselves, but rather how they change with Zo and x in TLM. The simplest possible Zl is therefore taken for the representation: purely resistive and greater than Zo as the most reasonable assumption (Fig. 4.12(b)). The line is given a length corresponding to the width of the monopole. In TLM, its impedance plot with frequency for a given Zo will be a circle. If Zo is 50 Ω, the magnitude of the reflection coefficient remains constant. The path taken in the Smith Chart with

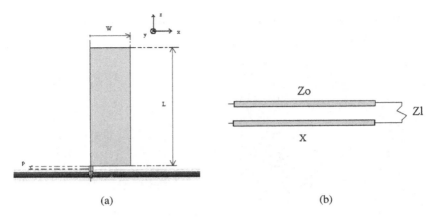

(a)   (b)

Fig. 4.12.   (a) Half of a conventional PMA and (b) TLM analysis.

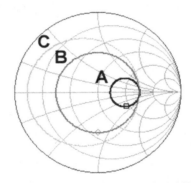

Fig. 4.13.   Smith Chart (0–10 GHz) for Fig. 4.12(b) with $Zl = 150\,\Omega$, $Zo\_1 = 100\,\Omega$ (black), $Zo\_2 = 50\,\Omega$ (grey) and $Zo\_3 = 25\,\Omega$ (light grey).

frequency is a clockwise turn or rotation that traces a circumference centred in the chart (curve B in Fig. 4.13). If Zo were greater or less than 50 Ω, the centre of the circumference would then drift on the horizontal axis, towards the right or the left, respectively, as shown by curves A and C in Fig. 4.13. The greater the difference between the load Zl and the line's Zo, the greater the radius of the circle and the more difficult it will be to achieve broadband matching. The intersection of the circumferences is the normalised load.

4.4.6.2. *Translation to antenna design*

In the simplified domain, TLM is useful for producing the same movements and loops that the PMA impedance plot requires in order to be matched (centred in

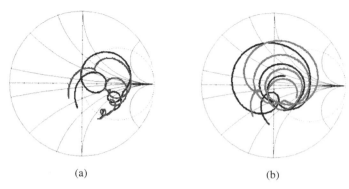

(a)   (b)

Fig. 4.14. Smith Charts (1–20 GHz) corresponding to widening the PMA in Fig. 4.12. (a) from $W = 19$ (black) $W = 28$ (grey) with $L = 45$ and $p = 3$ and (b) from $W = 19$ (black) to $W = 28$ (grey) with $L = 45$ and $p = 0.5$.

the chart). Figure 4.14 gives examples of these transformations. The effect of lengthening W in the PMA of Fig. 4.12(a) (lengthening x in Fig. 4.12(b)) is analysed for two heights p (different Zo) in figures (a) and (b), respectively. In both cases, when the monopole is widened, there is a clockwise rotation, analogous to the situation that occurs with a resistive load in TLM (Fig. 4.13). It is also observed how, in the first case, this rotation is less marked at high frequencies. This is consistent with the TLM model, since when p is greater (3 mm) and the frequency is also higher (20 GHz), the electrical distance with respect to the ground plane ceases to be small in electrical terms.

The change in p from 3 to 0.5 mm (reduction of Zo) makes the impedance plot expand over the chart as the black circle changed to the grey one in Fig. 4.13. This can be seen when curves of the same W length are compared in Fig. 4.14.

Due to the fact that the load used in the analogy for the TLM case is purely resistive and the one which it is attempted to be matched in the antenna's impedance chart is not only unknown but also arbitrary, the variations must be understood as relative to the original situation. The numerical values from which one starts in TLM do not therefore have a decisive importance. We will see the usefulness of this concept when designing a rectangular broadband antenna.

4.4.7. Case study 2: Broadband matching of a full-rectangular planar monopole case

The rectangular PMA antenna with centre feed can be understood to be the parallel of two antennas (lines) as in Fig. 4.12. As this is a parallel circuit, the impedances are reduced, moving towards the left of the Smith Chart (Fig. 4.15).

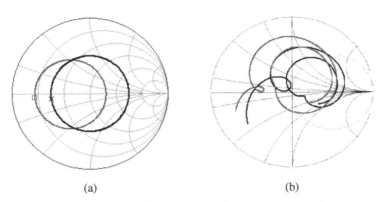

(a)   (b)

Fig. 4.15. Smith Chart (1–10 GHz) of semi (black) and full (grey) rectangular monopoles in accordance with (a) TLM or (b) PMA. Parameters for TLM: $Zl = 150\,\Omega$, $x = 30$ mm, $Zo = 25\,\Omega$. Parameters for antennas: $L = 45$ mm, $W = 30$ mm, $p = 1$ mm.

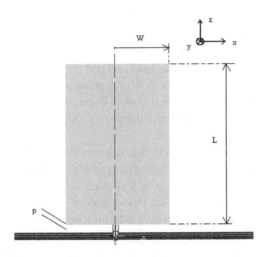

Fig. 4.16. Rectangular PMA to be optimised, with its corresponding parameters.

Suppose that, as a matching requirement, it is proposed to design a rectangular antenna with a lower limit (VSWR < 2) of 1.65 GHz and maximum bandwidth, while maintaining its rectangular shape, as shown in Fig. 4.16. No precise radiation conditions are defined.

The lower frequency principally defines the height above the ground plane. As a first approximation, the monopole is given a height L equal to $\lambda/4$ at that frequency above the ground plane for a linear monopole (45 mm). To ensure that this

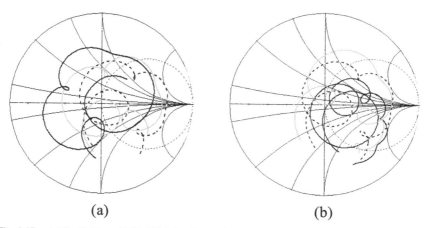

(a)   (b)

Fig. 4.17. (a) Smith Chart (1–20 GHz) for changes in distance p over the ground plane for a rectangular PMA. $W = 12$ mm, $L = 45$ mm, $p = 0.5$ mm (black), 1.75 mm (grey), 3 mm (dotted), (b) Smith Chart (1–20 GHz) with changes in the distance W for a rectangular PMA. $L = 45$ mm, $p = 3$ mm, $W = 6$ mm (black), 12 mm (dotted), 18 mm (grey).

specification is always complied with, the height L must be maintained the same. The problem reduces to changing W and p, the effects of which, in accordance with TLM, have already been studied on the Smith Chart.

Figure 4.17 shows the effects produced when the distance p above the ground plane is changed, for a fixed W of 12 mm (on the left), and when the width W is changed, for a fixed p of 3 mm (on the right). Following the criteria seen in the previous section, the optimised design is given for the case which is common to both, marked in dotted lines.

The bandwidth obtained, given by the condition VSWR < 2, is from 1.2 GHz to 3.8 GHz (ratio 3.17:1). The lowest frequency in the band has been reduced with respect to the planned value, due to the transverse dimension W, which gives a longer current path. However, this bandwidth is still not sufficient for UWB applications defined by the FCC (3.4:1). In the following chapters, more elaborate designs will be studied, based on the same techniques.

Chapter 5

UWB Monopole Antenna Bandwidth Synthesis

Daniel Valderas and Juan I. Sancho
CEIT and Tecnun, University of Navarra

Having seen how the transmission line model (TLM) can be applied to a monopole antenna, this chapter uses the model to synthesise the required bandwidth. PMAs are studied, because they are the simplest case, although the methodology can also be extrapolated to other types of antenna.

5.1. Introduction

At the end of the previous chapter, an example of rectangular PMA design was given. Despite obtaining a maximum bandwidth for that type of antenna through the use of TLM, this is not sufficient to cover the whole UWB spectrum as defined by the FCC (3.1–10.6 GHz) or for future applications requiring greater bandwidths.

One of the ways to increase the bandwidth is by giving the edge closest to the ground plane a staircase profile. Through the use of this technique, some designs can multiply fourfold the difference between the upper and lower frequencies of the bandwidth (Fig. 5.1) (Su *et al.*, 2004).

Certain notched designs result in a longer current path which, combined with coupled monopoles, results in a more compact PMA (Dou and Chia, 2000). Asymmetrical notches have also been satisfactorily introduced in designs using a coplanar waveguide (Liu, 2004). They have also been used successfully in PDA mobiles or mobile communications terminals covering the range from 1.65 to 10.6 GHz (Jung *et al.*, 2005). At high frequencies, the measured radiation pattern shows serious distortions in the horizontal plane. A complementary technique consists in making slots in the surface of the monopole.

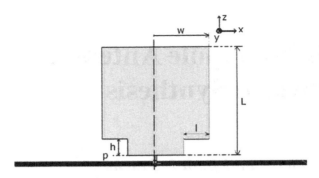

Fig. 5.1.   Square PMA with two notches in the bottom edge for increased bandwidth.

In the following sections, both techniques will be approached from a TLM perspective. However, neither of them defines the lower operating frequency limit, which must first be determined based on the length of the antenna's perimeter.

5.2. Defining the Lower Limit of the Frequency Band

To determine the lower frequency limit for a PMA such that VSWR < 2, experimental formulae are established by starting from the case of a monopole with a circular cross-section. In this case, the length corresponding to the resonant frequency is determined by Eq. (5.1) (Balanis, 1997).

$$L = 0.24 \cdot \lambda \cdot F, \tag{5.1}$$

where the factor F reflects the widening of the antenna due to the radius r, Eq. (5.2) is used.

$$F = \frac{L}{L+r}. \tag{5.2}$$

By combining both expressions, an analytical expression can be reached for the monopole's first resonant frequency (5.3).

$$\lambda = \frac{L}{0.24 \cdot F} = \frac{L+r}{0.24} = \frac{c}{f} \rightarrow f = \frac{0.24 \cdot c}{L+r}. \tag{5.3}$$

For the case where the antenna's cross-section is neither circular nor electrically small in all its dimensions, several solutions have been provided. Many use an equivalent radius r_e of an imaginary cylindrical antenna, based on the starting geometry. Thus, for triangular and square shapes with a side of length a, the equivalent radius would be $0.42a$ and $0.59a$, respectively (Hallen, 1938; Lo, 1953). There are corresponding relationships for other geometries (Balanis, 1997).

To calculate the equivalent radius r_e for a PMA, its area is considered to be equal to the area of the side of a cylindrical monopole of radius r_e. For example, for a rectangular PMA with sides $2W$ and L, we use Eq. (5.4) (Kumar and Ray, 2003).

$$2\Pi r_e L = 2WL \rightarrow r_e = \frac{W}{\pi}. \tag{5.4}$$

Applying the equivalent radius concept of a PMA is valid for predicting the lower limit of the band for the PMA (f_m) for VSWR < 2, by using it to replace r in Eq. (5.3). The fact that the monopole is widened reduces this frequency, allowing a slight reduction in the height of the antenna for the same frequency operation. Equation (5.5) also includes the length of the feed, p.

$$f_m = \frac{0.24 \cdot c}{L + r_e + p}. \tag{5.5}$$

5.3. Obtaining the Upper Frequency with Staircase Profile in TLM

5.3.1. *One step in a PMA's profile according to TLM*

Figure 5.2 shows (a) a line loaded with a resistive impedance and (b) two lines, half as long, with different characteristic impedances, carrying the same resistive load (a transformer with two equal sections). By changing to two lines that are half as long, a new lobe is added to the corresponding impedance chart in Fig. 5.2(c). Point A corresponds to each complete turn in the Smith Chart (each increase of $\lambda/2$ in (a) in the figure — the black trace). Point B, in the centre of the new lobe, is obtained for every odd multiple of $\lambda/2$ in (b). In this way, the odd resonances are accentuated and the even ones remain unchanged (Fig. 5.2(d)). The pattern is therefore repeated at every second turn in the Smith Chart: every time that the total length of the line is a multiple of λ. With the dimensions given in the example, this occurs for each 10 GHz increment ($x = 30$ mm).

By taking other parameters Zo_1, Zo_2 and Zl, the size and location of the lobe is changed, but it is always added to the initial impedance chart at the frequency that corresponds to $x = \lambda/2$ which would be an identifiable point for the antenna where $W = \lambda/2$.

Now it is a question of confirming the parallel situation in a rectangular PMA. This consists in making a cut or notch in the profile of the lower edge, as shown in Fig. 5.3. The effects on the Smith Chart are presented in Fig. 5.4. Here we can see that when a step-shaped notch is made in the monopole, a lobe is added and point A

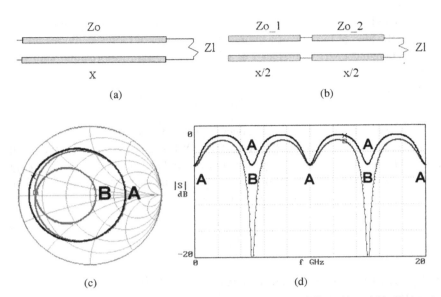

Fig. 5.2.   Passing from a line with constant Zo (black) to a stepped line with variable Zo (grey). ($Zl = 150\,\Omega$, $x = 30$ mm, $Zo = Zo\_1 = 25\,\Omega$, $Zo\_2 = 40\,\Omega$.)

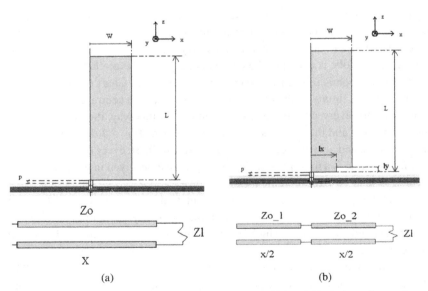

Fig. 5.3.   Notch implemented in a PMA and the corresponding TLM model in the form of a transformer having two sections.

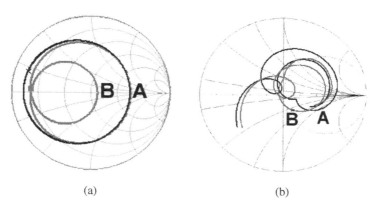

Fig. 5.4. Smith Chart (1–10 GHz) of the cases described in Fig. 5.3 for the (a) TLM and (b) PMA domains. The black trace is for Fig. 5.3(a); the grey for Fig. 5.3(b). Points A and B correspond to $W = \lambda/2$. TLM parameters: $Zl = 150\,\Omega$, $x = 30$ mm, $Zo\_1 = 25\,\Omega$, $Zo\_2 = 40\,\Omega$. PMA parameters: $L = 45$ mm, $W = 30$ mm, $p = 1$ mm, $lx = 15$ mm, $ly = 2$ mm.

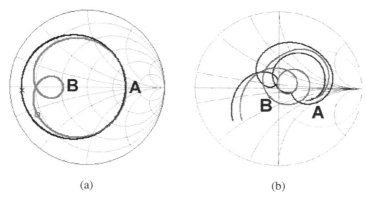

Fig. 5.5. Smith Chart (1–10 GHz) of the cases described in Fig. 5.3 for the (a) TLM and (b) PMA domains. The black trace is for Fig. 5.3(a); the grey for Fig. 5.3(b). Points A and B correspond to $W = \lambda/2$. TLM parameters: $Zl = 150\,\Omega$, $x = 30$ mm, $Zo\_1 = 25\,\Omega$, $Zo\_2 = 60\,\Omega$. Antenna parameters: $L = 45$ mm, $W = 30$ mm, $p = 1$ mm, $lx = 15$ mm, $ly = 4$ mm.

of the rectangular PMA is converted into point B of the notched PMA. Both correspond to the frequency at which the width of the monopole is $\lambda/2$ (5 GHz in the example).

If we increase the characteristic impedance Zo_2 (Fig. 5.5(a)), the size of the new lobe is reduced. In the same way, if the notch in the rectangular PMA, ly in Fig. 5.3(b), is made higher, the added lobe tends to turn back on itself more (Fig. 5.5(b)).

In this way, the effect of notches in an antenna is described in TLM by considering transmission lines having variable characteristic impedance, which are equivalent to a multi-section impedance transformer. This translates into new lobes in the Smith Chart. When the notches are abrupt, at right angles, the new lobes are added at set frequencies. As will be shown below, these allow an upper limit for the bandwidth to be clearly established. This is defined by the last lobe introduced by the staircase profile. Intuitively, the more steps there are in the profile, the greater the number of resonances introduced and the greater the bandwidth.

5.3.2. *Two steps in a PMA's profile according to TLM*

In TLM, for the case of a transformer having three sections, the Smith Chart, and the reflection coefficient $|\Gamma_{11}|$ are represented in Fig. 5.6.

In the corresponding impedance plot in figure (c) two new lobes are added, so that point A in the black trace is analogous to point B (one turn at $\lambda/2$) and to point C (two turns at $\lambda/2$) in the grey trace. At the third turn, both plots coincide. This phenomenon is cyclical, it is repeated every three turns in the Smith Chart of the initial situation, i.e., every time the total length of the line is a multiple of $3\lambda/2$. With the dimensions of the example, this occurs for every increment of 15 GHz in TLM.

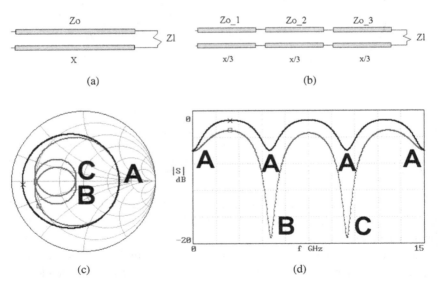

Fig. 5.6.   Passing from a line with constant Zo (black) to a line with variable Zo, having two steps (grey). $Zl = 150 \, \Omega$, $x = 30$ mm, $Zo\_1 = 25 \, \Omega$, $Zo\_2 = 40 \, \Omega$. $Zo\_3 = 60 \, \Omega$.

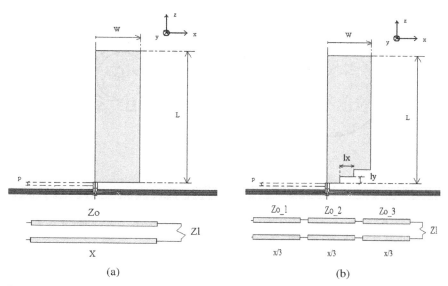

Fig. 5.7. Steps implemented in PMA and the corresponding TLM in the form of a three-section transformer.

Figure 5.7 shows this in the PMA. The results are represented in Fig. 5.8. Here, one can see how, when two equal steps are introduced in the monopole, two lobes are added: one (the larger) in point B and another at point C. These points correspond to the frequencies at which the width of the monopole is $\lambda/2$ (5 GHz for the example) and λ (10 GHz). Both points coincide, in terms of frequency, with the respective points in Fig. 5.6.

5.3.3. *Analytical estimate of the upper limit of the band for a rectangular staircase monopole*

In a general case, like the one shown in Fig. 5.9, one can see how, as N notches or steps are made at both sides of the feed, N new resonances (A, B and C) are introduced, located every 5 GHz, which is the frequency at which $W = \lambda/2$. Thus, A is found at 5 GHz, B at 10 GHz and C at 15 GHz. When more notches are added, the number of added resonances increases, but the spacing of their frequencies is set by the width of the antenna (5 GHz for 30 mm). The frequency condition for the added lobes for the generic case is expressed by Eq. (5.6) where c is 3E8 m/s (Valderas *et al.*, 2007a).

$$W = n\frac{\lambda_n}{2} \longrightarrow f_n = \frac{nc}{2W}, \quad n = 1, 2, \ldots, N \qquad (5.6)$$

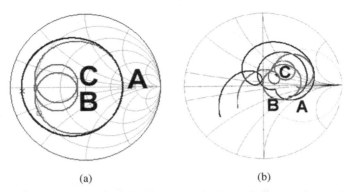

(a)   (b)

Fig. 5.8.   Smith Chart (1–15 GHz) of the cases described in Fig. 5.7 for the (a) TLM and (b) PMA domains. The black trace is for Fig. 5.7(a); the grey for Fig. 5.7(b). TLM parameters: $Zl = 150\,\Omega$, $x = 30$ mm, $Zo\_1 = 25\,\Omega$, $Zo\_2 = 40\,\Omega$ and $Zo\_3 = 60\,\Omega$. PMA parameters: $L = 45$ mm, $W = 30$ mm, $p = 1$ mm, $lx = 10$ mm, $ly = 2$ mm.

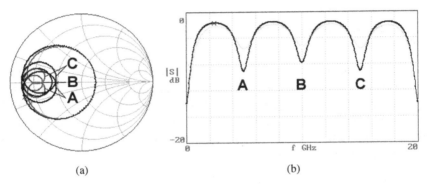

(a)   (b)

Fig. 5.9.   (a) Smith Chart (1–20 GHz) and (b) $|\Gamma 11|$ for a four-section transformer ($Zl = 150\,\Omega$, $x = 30$ mm). In black: $Zo\_1 = 25\,\Omega$, $Zo\_2 = 37\,\Omega$, $Zo\_3 = 47\,\Omega$, $Zo\_4 = 60\,\Omega$. In grey: $Zo\_1 = 25\,\Omega$, $Zo\_2 = 43\,\Omega$, $Zo\_3 = 61\,\Omega$, $Zo\_4 = 80\,\Omega$.

However, if the jump in Zo between two consecutive steps is more abrupt (Fig. 5.9 in grey), the lobes tend to decrease in size. This is a way of controlling the added resonances.

A bevelled cut may be considered to be an extreme case of the staircase profile ($N \to \infty$). A more or less sharply sloping bevelled profile corresponds to a larger or smaller jump between steps (Fig. 5.9).

The cases in which two or three steps are made in both sides of a PMA are shown below (PMA-2N in Fig. 5.10 and PMA-3N in Fig. 5.11). In the first case, having two steps, there are two new lobes: at 5 GHz and at 10 GHz, if W is

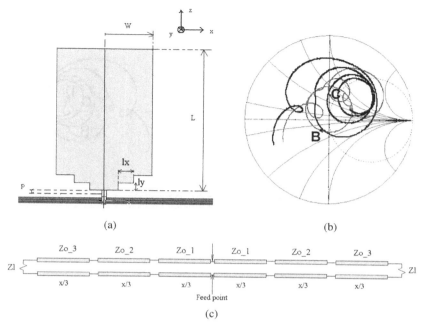

Fig. 5.10. (a) Making two notches at each side (PMA-2N), (b) Smith Chart (grey) (1–20 GHz) together with the one for a rectangular PMA (black), (c) TLM model of PMA-2N $L = 45$ mm, $W = 30$ mm, $p = 1$ mm, $lx = 10$ mm, $ly = 2$ mm.

30 mm. In the second case, with three steps, there are three new lobes: at 5, 10 and 15 GHz.

In this way, once the resonances have been introduced, they can be centred in the chart by modifying the parameters ly, p and W (as has been explained in the previous chapter), to give the antenna with the required bandwidth.

A formula could therefore be established which would serve as a guide for the upper limit of the bandwidth for a PMA with a staircase profile, for the condition VSWR < 2. Thus, Eq. (5.7), based on the condition in Eq. (5.6), would allow pairs of values N and W to define the said limit (Valderas *et al.*, 2007a).

$$f_u \approx N \frac{0.3}{2W}, \tag{5.7}$$

where f_u is the upper frequency of the band, in GHz, N is the number of pair of notches or steps and W is half the width of the monopole, in metres.

As a general rule, for a given width W, the greater the number of notches or steps, the greater the bandwidth. Another way of achieving the same objective

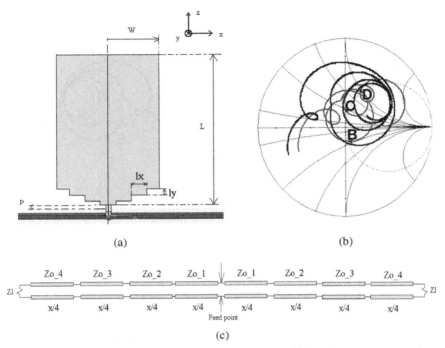

(a)   (b)

(c)

Fig. 5.11.   (a) Making three notches at each side (PMA-3N), (b) Smith Chart (grey) (1–20 GHz) together with the one for a rectangular PMA (black) (©2007 IEEE), (c) TLM model of PMA-2N. $L = 45$ mm, $W = 30$ mm, $p = 1$ mm, $lx = 7.5$ mm, $ly = 4/3$ mm.

would be to reduce the width. However, obviously, the condition VSWR < 2 could not be maintained below a certain W threshold value: the antenna would become a multifrequency one.

The sharp cut of a step, compared to the smoothed-out cuts of a bevel allows the resonances to be clearly set by the discontinuities in the profile and allows the upper limit of the band to be determined. As mentioned in Chap. 2, the antenna acts as a band-pass filter; a difficult effect to obtain in the case of rounded cuts or a bevelled shape.

5.4. Obtaining the Upper Frequency through Slot Etching

Figure 5.12(a) shows a PMA in which a horizontal slot has been implemented, at height H. Figure 5.12(b) shows the effect on the Smith Chart for changes in H. It is clear that the slot adds a new resonance, which becomes more pronounced the closer the slot is to the zone equivalent to the transmission line.

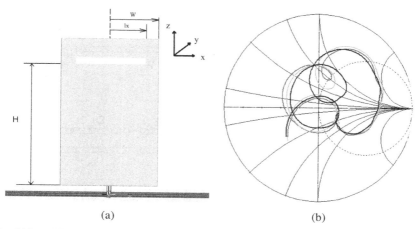

(a) (b)

Fig. 5.12. (a) Rectangular PMA $W = 14.5$ mm, $L = 45$ mm, $p = 2.7$ mm with a horizontal slot $lx = 12$ mm, (b) Smith Chart (1–12 GHz) for $H = 10.3$ mm (grey), $H = 7.3$ mm (light grey), $H = 4.3$ mm (black).

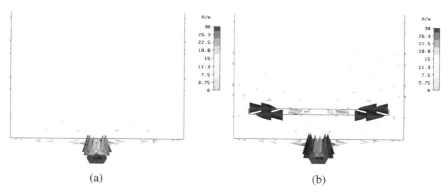

(a) (b)

Fig. 5.13. Current density distribution at 6.25 GHz for the PMA in Fig. 5.12: (a) with no slot and (b) with a slot at $H = 7$ mm.

Figure 5.13 shows the current density distribution for the PMA without a slot and with a slot at a small height H, at the same phase. In the second case, it can be seen that the current circulates in opposite directions at the opposite sides of the slot. In this way, **a new transmission line is introduced,** located between the original transmission line of the model and the load Zl or radiating structure.

Figure 5.14 shows the corresponding representations in TLM for the cases with and without a slot.

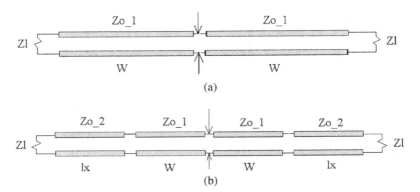

Fig. 5.14.   TLM representation of the rectangular PMA, (a) with no slot and (b) with a slot.

Compared to the case of the staircase shape, in which the first new resonance is introduced at a frequency such that the value of W is $\lambda/2$, the new condition now changes to Eq. (5.8).

$$W + lx = \lambda/2. \tag{5.8}$$

So the frequency of the new lobe has been reduced, depending on the parameter lx for the same width W.

Figure 5.15(a) shows the Smith Chart for Fig. 5.14. The fact that a step has been added is shown by a new resonance. This also occurs when the cases with

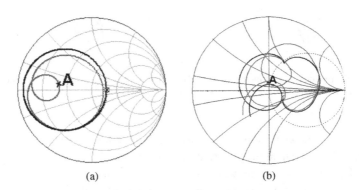

Fig. 5.15.   Smith Chart (1–12 GHz) for Fig. 5.14 (left) and for the antenna in Fig. 5.12(a) (right), without a slot (black) and with a slot (grey). $W = 14.5$ mm, $L = 45$ mm, $p = 2.7$ mm, $lx = 12$ mm, $H = 4.3$ mm, width of slot $= 1$ mm. Point A at 5.7 GHz.

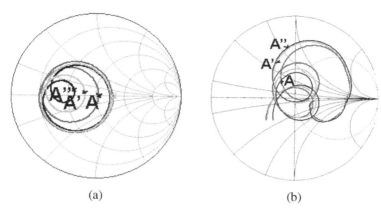

(a)   (b)

Fig. 5.16. (a) The effect on the Smith Chart (1–12 GHz) of changes in Zo_2 in Fig. 5.14 and (b) of the width of the slot in a PMA. Antenna parameters: $W = 14.5$ mm, $L = 45$ mm, $p = 2.7$ mm, $lx = 12$ mm, $H = 4.3$ mm. Variable width of slot: 2 mm (grey), 4 mm (light grey) and 6 mm (black). Line parameters: $Zl = 150\,\Omega$, $Zo\_1 = 50\,\Omega$, Zo_2 variable $= 60\,\Omega$ (grey), $70\,\Omega$ (light grey) and $80\,\Omega$ (black).

and without a slot are compared for the antenna in Fig. 5.12(a). In them, point A fulfils the condition of Eq. (5.8).

The size of the lobe is controlled by the width of the slot. This is analogous to the way that the size of the lobe is controlled by the height of a step in a staircase profile. If they are increased, the characteristic impedance is also increased. The same changes in the lobe are shown for both TLM and PMA, in Fig. 5.16.

In addtion to the increase in Zo_2 when the width of the slot is increased, the frequency at which the lobe is added is slightly reduced, as the effective length of the equivalent line is increased. This places points A, A′ and A″ in Fig. 5.16(b) further from the lobe as the width increases. These points occur at 5.66 GHz, which is the frequency that fulfils the condition in Eq. (5.8) in this case.

Seen from another point of view, the slot could be considered, not just as a device to produce a resonance through a new transmission line, but as a superimposed radiating element, which would present its resonance according to the condition in Eq. (5.9).

$$2lx = \lambda/2. \tag{5.9}$$

This implies a condition that is quite close to Eq. (5.8), as lx is approximately W. In this way, two antennas would be implemented in one. From this point of view, it is logical that the slot has its effect in the area in which it is excited, i.e., where there is a high density of horizontal current at small distances H.

Three examples that illustrate the above techniques are now presented. For simplicity, rectangular PMAs will be used. It will be demonstrated that, for the same dimensions of the outline shape, different bandwidths can be obtained, not only UWB (3.4:1) but also, e.g., 8.6:1 and 14.1. In all cases, a lower reference frequency of 1.4 GHz will be taken (below the 3.1 GHz given for FCC UWB). This is so that the band's upper frequency (for the same ratio between the ends of the band) is not too high, for convenience when making measurements. This can be adjusted to the required frequencies by appropriate scaling.

5.5. Case Study 1

5.5.1. *Design*

The required values are: a lower band limit that is slightly below 1.4 GHz and a bandwidth ratio (between the ends of the band) of 3.4:1. The steps for designing a slotted PMA (PMA-S) will be given.

The starting point is a rectangular PMA. Its dimensions L, W and p are defined by the lower frequency. In this case, the values taken are 45, 14.5 and 2.7 mm, respectively (Eq. (5.5)).

The size of the ground plane is 260 mm $\times$ 260 mm, as for all the designs in this book. This means that the electrical length of the side is 1.2 λ, for the common lower frequency of 1.4 GHz. This size is such that the impedance measurement tends to be independent of the ground plane (Arai (2001), pp. 41–43). If the ground plane's size is reduced below a critical value, the lower limit of the band will rise. This allows one to concentrate on the design of the monopole itself.

The slot is designed so that its length fulfils the conditions described above at a frequency close to the designed upper limit (e.g., 4.76 GHz). A total length of 36 mm provides a resonance at approximately 4.61 GHz, in accordance with the condition in Eq. (5.8), and at 4.16 GHz, in accordance with the condition in Eq. (5.9). Since, in this case, the slot must have a length lx that is greater than W, the slot must be U-shaped (Fig. 5.17).

The value given to lx is 10 mm and to ly is 8 mm. The resulting antenna is shown in Fig. 5.18. The slot is made 5 mm above the lower edge of the monopole. The width e of the slot is 2 mm.

The design parameters corresponding to the end of the design process are shown in Table 5.1. The width of the soldering pin is 1.2 mm.

In this design, the vertical section of the slot has been kept as short, i.e., as low as possible. Otherwise, it would be close to the vertical limit of the antenna which, in electrical terms, this would mean an open circuit. In these circumstances, at the

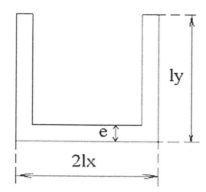

Fig. 5.17.   Parameters of the slot to be implemented.

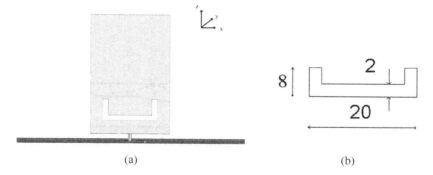

(a)   (b)

Fig. 5.18.   Configuration of the PMA-S and the implemented slot.

frequency at which Eq. (5.10) is fulfilled, it would behave like an open circuit to short circuit transformer and the desired effect would not be achieved.

$$lx + ly = \lambda/4. \tag{5.10}$$

5.5.2. *Simulation and measurements*

5.5.2.1. *Impedance bandwidth*

Figure 5.19 shows the measured bandwidth for this prototype, compared with the same antenna with no slot, i.e., a purely rectangular antenna (Valderas *et al.*, 2005).

Table 5.2 shows the two ends of the band for both cases and how the slotted antenna allows the UWB specifications to be fulfilled for the VSWR < 2 condition.

Table 5.1.   Parameters of the PMA-S (mm).

L	W	p	e	lx + ly	H
45	14.5	2.7	2	18	5

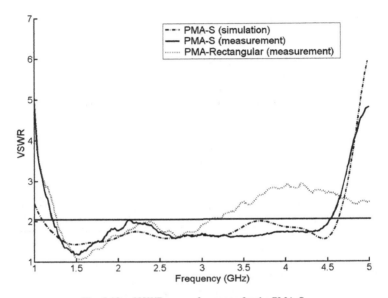

Fig. 5.19.   VSWR versus frequency for the PMA-S.

Table 5.2.   Impedance bandwidth for the rectangular PMA with and without a slot.

	Lower limit (GHz)	Upper limit (GHz)	Bandwidth	UWB Requirement (3.4:1)
Rectangular PMA	1.25	3.1	2.5:1	X
PMA-S	1.2	4.5	3.75:1	√

5.5.2.2. *Radiation patterns at representative frequencies*

Figures 5.20 to 5.22 show the measured radiation patterns for the PMA-S at the three principal cuts and at representative frequencies in the interval (1.2, 2.6 and 4.5 GHz) (Valderas *et al.*, 2005). The corresponding axes are shown in Fig. 5.18. These diagrams are very similar to those for other PMAs.

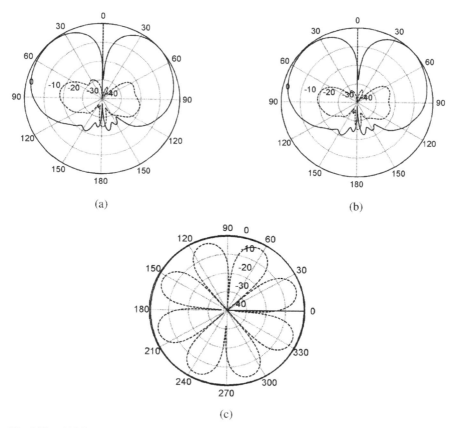

(a)

(b)

(c)

Fig. 5.20.   (a) Measured radiation patterns for the PMA-S at 1.2 GHz in the $\Phi = 0°$ plane, (b) in the $\Phi = 90°$ plane and (c) in the $\theta = 90°$ plane. Co (-) and cross (- -) polarisation.

It can be seen that the type of pattern is typically monopolar, with a null in the z direction and a predominantly copolar E_θ component. The ground plane limits the radiation in the $-z$ direction, but does not eliminate it, because it is finite.

The patterns keep this general appearance over frequency, although the maximum drifts in the vertical cuts, but with a reasonable level of stability. In these cuts, the cross-polarisation is far less than the co-polarisation, due to the fact that the antenna is planar.

We can see that over this range (3.4:1), the patterns are sufficiently stable. It is not necessary to study the pattern at other intermediate frequencies to be able to state that there is 100% uniformity in the H plane over this range of frequencies. Omnidirectionality and matching have good characteristics between similar limits.

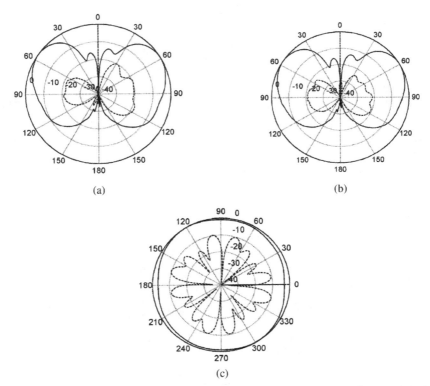

Fig. 5.21.   (a) Measured radiation patterns for the PMA-S at 2.6 GHz in the $\Phi = 0°$ plane, (b) in the $\Phi = 90°$ plane and (c) in the $\theta = 90°$ plane. Co (-) and cross (- -) polarisation.

A description of the general stability of the diagrams would not be complete without analysing the antenna's absolute gain, which is shown in Fig. 5.23. Between 1.4 and 4.2 GHz, the gain is around 5 dB, varying by ± 1 dB.

5.6. Case Study 2

The requirement is for a PMA with the same lower operating limit (1.4 GHz) and more than twice the bandwidth of the previous case (8.6:1 — upper limit 12 GHz). Essentially the same dimensions are maintained for the perimeter of the antenna.

5.6.1. Design

If a single notch is made at each side (PMA-1N), as shown in Fig. 5.24, the added resonances will correspond to $W = \lambda/2$, $W = 3\lambda/2$ and so on. The width of the

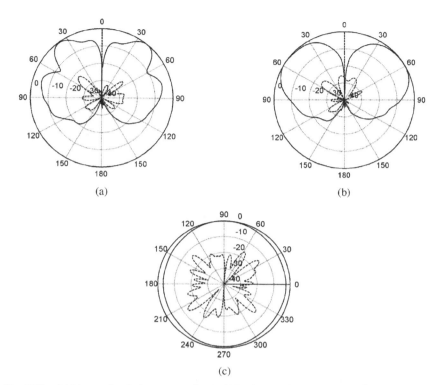

Fig. 5.22.   (a) Measured radiation patterns for the PMA-S at 4.5 GHz in the $\Phi = 0°$ plane, (b) in the $\Phi = 90°$ plane and (c) in the $\theta = 90°$ plane. Co (-) and cross (- -) polarisation.

Table 5.3.   Parameters of the PMA-1N (mm).

L	W	p	lx	ly
45	14	1.4	7	4.2

monopole will have to be such that W is approximately $\lambda/2$ at 12 GHz (12.5 mm). The dimensions of the prototype are shown in Table 5.3.

5.6.2. *Simulation and measurements*

Figure 5.25 shows the VSWR for the PMA-1N.

It is observed that the measured bandwidth covers a range between 1.3 and 11.5 GHz with VSWR < 2. The ratio between these frequencies is 8.8:1, which is larger than the required value (8.6:1).

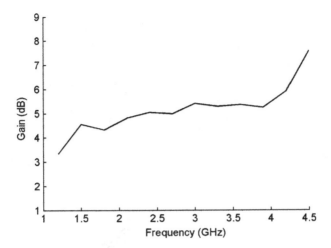

Fig. 5.23.   Maximum calculated gain (dB) for the PMA-S.

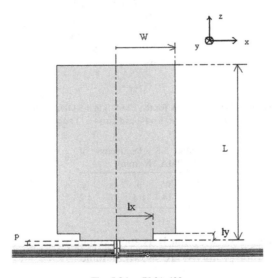

Fig. 5.24.   PMA-1N.

5.7. Case Study 3

The requirement is for a PMA having the same perimeter as in the previous study, but with a bandwidth of 14:1 (1.4–19.6 GHz). It is therefore necessary to use a larger number of notches.

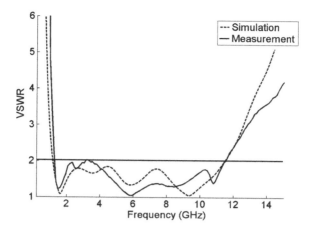

Fig. 5.25. VSWR versus frequency for the PMA-1N.

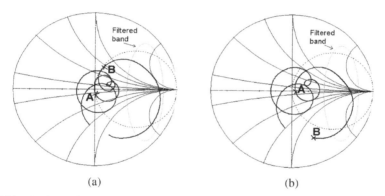

Fig. 5.26. (a) Smith Chart (1–30 GHz) for a PNA-2N and (b) PMA-1N with the same L and W dimensions. Points A and B at 10 and 20 GHz.

5.7.1. *Design*

As mentioned above, if PMA-2N is implemented (Fig. 5.27), the added frequencies are found where the value of W is $\lambda/2$ and λ. The antenna must be given a parameter W of approximately 15 mm, so that when two notches are made, the resonances are added at 10 GHz (for the condition $W = \lambda/2$) and 20 GHz ($W = \lambda$). Points A and B indicate these frequencies, which determine the added lobes (Fig. 5.26(a)).

Table 5.4 shows the dimensions of the PMA-2N (Fig. 5.27) that offers the specified operating range, with the same values for L and W as the PMA-1N of the previous section.

Table 5.4. Parameters of the PMA-2N (mm).

L	W	p	lx	ly
45	14	1	4.7	5

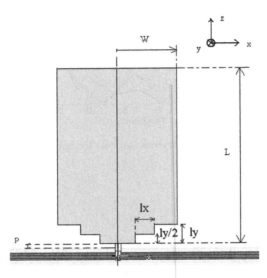

Fig. 5.27. PMA-2N.

Comparing Fig. 5.26(a) (PMA-2N) and 5.26(b) (PMA-1N with the same L and W dimensions as PMA-2N), the difference between the two cases is clear when the aim is to use the antenna as a filter: point B (corresponding to $W = \lambda$) does correspond to a resonance for the PMA-2N, but not for the PMA-1N.

If, instead of this profile, a bevel of equivalent dimensions had been made (with the same height $z = p + ly$ at $x = W$), although new lobes would be created (points A and B are also at 10 and 20 GHz), the required objective would not have been achieved (Fig. 5.28).

5.7.2. *Simulation and measurements*

Figure 5.29 shows the VSWR for the PMA-2N. It is observed that the measured bandwidth covers a range between 1.3 and 20.4 GHz with VSWR < 2. The initial specification is therefore satisfied.

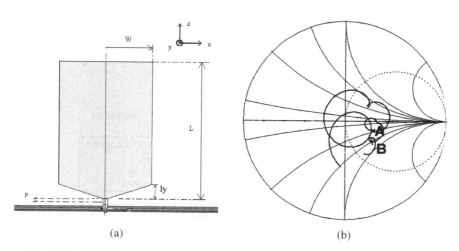

(a)　　　　　　　　　　(b)

Fig. 5.28.　(a) PMA with a bevel equivalent to PMA-2N and (b) the Smith Chart (1–30 GHz).

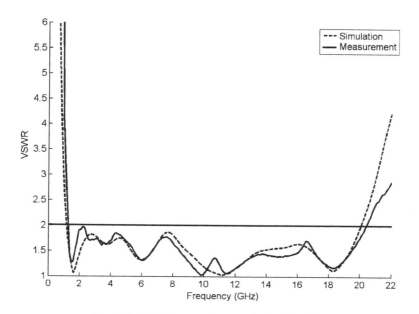

Fig. 5.29.　VSWR versus frequency for the PMA-2N.

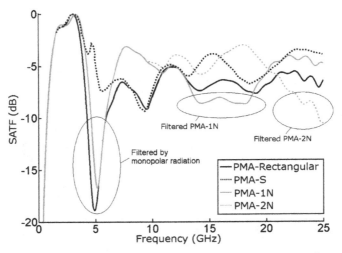

Fig. 5.30. SATF (dB) for the θ component in the H plane and at 1350 mm versus frequency for the PMAs seen in the chapter, in comparison with the purely rectangular antenna.

5.7.3. *Discussion: Impedance matching and transfer function*

In Chap. 2, the SATF parameter was introduced as representing, for the antenna, a transfer function between the input signal to the antenna and the pulse radiated in a UWB transmission. This section covers the impact of bandwidth control on the SATF parameter and the advantages that this provides.

Figure 5.30 shows the magnitudes of the SATF parameter in the H plane for the antennas considered in this chapter. The rectangular PMA, in which the steps were cut for the PMA-Ns and the slot was made for the PMA-S, is taken as the reference. The SAFT is calculated by averaging the transfer functions at $\Phi = -90°$, $-75°$, $-60°$, etc. The radiated electrical field E_θ is calculated in transmission at 1350 mm under far-field conditions, as a response to a Gaussian pulse.

It is observed how, around 5 GHz, for the PMA-Rectangular, PMA-1N and PMA-2N, filtering occurs in this plane, associated with the drop in gain. This arises from the radiation null in the H plane experienced by a linear monopole at the frequency at which its height is λ. If the total current path, given by Eq. (5.11), is taken into account, the frequency at which this occurs is exactly 5 GHz for the designs under consideration, in which the values of L and W were common.

$$W + L = \lambda. \tag{5.11}$$

If the plane under consideration is not the horizontal plane ($\theta < 90°$), the null which here occurs at 5 GHz will change in frequency. This filtering effect is useful if the direction of the transmission is known. This frequency can be controlled by adjusting the antenna's height and width.

The PMA-S should be considered separately. In this case, one observes how the radiation null at 5 GHz is cancelled out in the horizontal plane. This is due to the fact that the slot resonates like a superimposed antenna at that frequency. Its length is $\lambda/2$ at 4.16 GHz. As the slot's radiated component is fundamentally θ, the radiation null is cancelled out in the H plane.

Figure 5.30 also includes other filtering ways associated with the design of a UWB antenna based on TLM. The traces indicated obey the upper limit of the impedance bandwidth that has been controlled by the method (11.5 GHz for the PMA-1N and 20.4 GHz for the PMA-2N). Although this frequency filtering is not very pronounced, it presents two fundamental advantages. The first advantage is that it can be controlled independently of the null defined by condition (5.11). The second advantage is that filtering occurs equally for all spatial directions.

Thus, among all the possible applications of controlling a PMA's impedance bandwidth, the use of the antenna as a filter stands out for UWB transmission/reception. This would provide several advantages such as improving the system's immunity to undesirable interferences from any spatial direction, adjusting the antenna's transfer function to the pulse to be transmitted, complying with given emission limits, and reducing the requirements for RF elements in the circuit.

Chapter 6

UWB Monopole Antenna Bandwidth Maximisation

Daniel Valderas and Juan I. Sancho

CEIT and Tecnun, University of Navarra

Having analysed the transmission line model and how it can be used to define bandwidth through the use of a staircase profile and slots, we now set out the procedures for designing monopole antennas with the greatest possible impedance bandwidth. For simplicity, special radiation requirements will not be included (omnidirectionality or directivity), so we will work with planar structures over a perpendicular ground plane. However, the design techniques covered here are also applicable when these requirements are present.

The chapter ends with a study of the repercussions of a high bandwidth on spectral efficiency in comparison with other monopole antennas.

6.1. Introduction

One of the most extensively used techniques for increasing the bandwidth of a rectangular PMA is to make a bevelled notch or cut (Fig. 6.1). This technique allows the upper limit of the bandwidth to be modified by changing the angle of the cut: an antenna's bandwidth ratio can change between the upper and lower limits from 2:1 (VSWR < 2) for the initial design of the rectangular monopole, to 6:1 for an optimal design, with hardly any effect on the radiation pattern (Ammann, 2001).

As an analogous technique, rounded lower profiles are used (Fig. 6.2). In this respect, the use of a semicircular profile permits bandwidth ratio values of up to 7.69:1 to be reached (Anob *et al.*, 2001).

Both these modifications are in accordance with the model described in the previous chapter. It is no surprise that the principal cause of wide bandwidth in a PMA is the shape of the part closest to the ground plane.

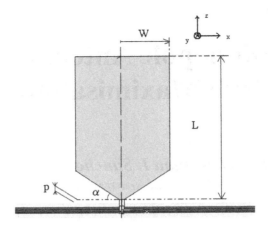

Fig. 6.1.   PMA with bevel-shaped lower edges.

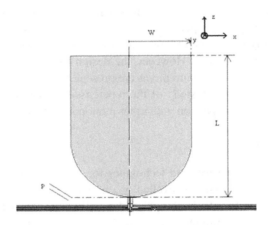

Fig. 6.2.   PMA with semicircular base.

Furthermore, when the current path along the perimeter is lengthened, the lower operating limit is reduced (Fig. 6.3) (Suh *et al.*, 2004). In the example here, a more compact antenna design is achieved compared to a disk-shaped monopole antenna, giving the same effect as increasing the height of the antenna above the ground plane.

Having established this lower limit, the modifications achieved by applying TLM to the lower profile will affect the bandwidth defined. If the requirement is

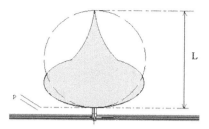

Fig. 6.3.   Planar Inverted Cone Antenna (PICA) (Suh *et al.*, 2004).

for the maximum possible bandwidth, smooth changes in the shape of the profile
are preferable to abrupt cuts.

6.1.1. *Modifying the profile of the edge of the PMA closest to the ground plane*

Although the characteristic impedance of the transmission line in the TLM model
is unknown, its relative changes can be estimated for each chosen profile. In order
to show this, in Fig. 6.4 three monopoles with different profiles are proposed with
common heights above the ground plane, p at $x = 0$ and $p + l_y$ at $x = W$.

Figure 6.5 shows a profile comparison of the lower edges of the three antennas.
Equations (6.1), (6.2) and (6.3) reflect how the heights y change with distance x
where p, k_1, k_2 and k_3 are determined by the initial and final values of the height
over the ground plane.[a]

$$y_a = p + (k_1 x)^2, \tag{6.1}$$

$$y_b = p + k_2 x, \tag{6.2}$$

$$y_c = p + (k_3 x)^{\frac{1}{2}}. \tag{6.3}$$

6.1.2. *Applying TLM: Changing the characteristic impedance*

According to Eq. (4.1), the expression relating the height of the conductor above the
ground plane to its characteristic impedance Zo is logarithmic.[b] Having normalised
the characteristic impedances by the maximum value of that expression (which is

---

[a]In Fig. 6.5. Profiles corresponding to Fig. 6.4, when $x = 0$, $y = p = 1$ mm and when $x = 30$ mm,
$y = p + l_y = 11$ mm.
[b]The value of 60 mm is taken as an estimated figure for the cross-section of antenna through which the
horizontal current passes.

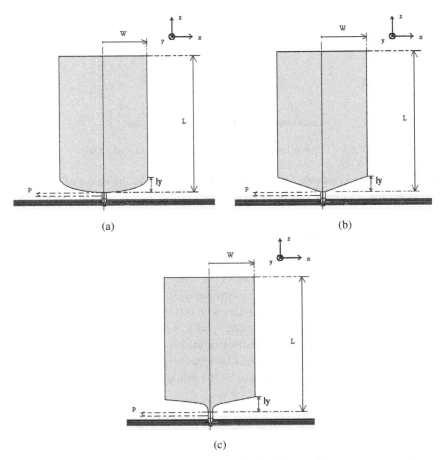

Fig. 6.4.   Lower edge profiles: (a) convex, (b) bevelled and (c) concave. The common parameters are: $L = 45$ mm, $W = 30$ mm, $ly = 10$ mm, $p = 1$ mm.

common to the three cases for the same final height of 11 mm), the graphs of the characteristic impedances corresponding to the three antennas would be as shown in Fig. 6.6.

This situation may be qualitatively approximated by a stepped series of transmission lines having discrete values for Zo. Taking a five-section transformer like that shown in Fig. 6.7 with the parameters shown in Table 6.1, the relative movements in the Smith Chart for a purely resistive load are represented in Fig. 6.8(a). One observes that there is the same number of new lobes or resonances in the three cases, but the separation between them differs.

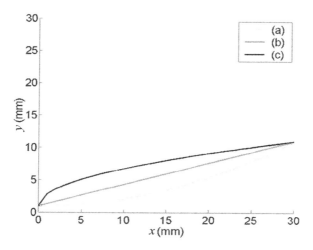

Fig. 6.5.   Profiles corresponding to Fig. 6.4.

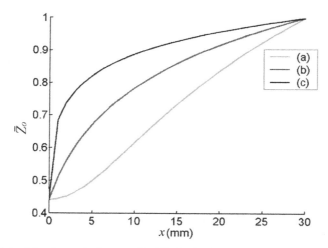

Fig. 6.6.   Plots of the changes in the normalised Zo for the three profiles in Fig. 6.5 according to the formula $Zo = 60 \ln \frac{4h}{d}$.

Analogously, Fig 6.8(b) shows a parallel effect on PMAs having the profiles shown in Fig. 6.4. The first lobe, at the lowest frequency in the three cases, is common to the initial rectangular planar structure (marked as A in Fig. 6.8(b)). The other resonances are added by the new curved profile and correspond to those in Fig. 6.8(a). This corresponds in qualitative terms to the relative movements in

Fig. 6.7.   TLM approximation in five steps to the continuously variable profile.

Table 6.1.   Values assigned to the Zo for TLM in Fig. 6.7 for a qualitative analysis of the PMA profiles in Fig. 6.4.

	Profile (a)	Profile (b)	Profile (c)
Zo_1 (Ω)	50	50	50
Zo_2 (Ω)	80	105	130
Zo_3 (Ω)	110	135	150
Zo_4 (Ω)	140	155	165
Zo_5 (Ω)	170	170	170

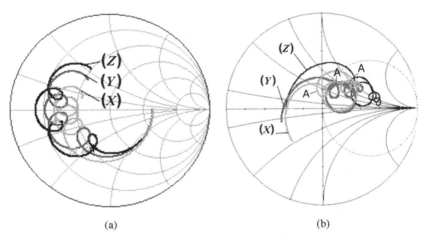

(a)   (b)

Fig. 6.8.   (a) Smith Chart (1–20 GHz) for the stepped profile using TLM shown in Fig. 6.7 with the parameters given in Table 6.1. $W = 30$ mm and $Zl = 250\ \Omega$, (b) Smith Chart (1–20 GHz) for the PMAs in Fig. 6.4 with $ly = 10$ mm.

the Smith Chart. The resonances introduced by the concave profile are the least favourable for achieving a high bandwidth, as they are the most widely separated (case Z). It is therefore not surprising that profiles that are convex or bevelled are chosen as more advantageous in the examples given at the start of the chapter, since they introduce resonances that may overlap in the Smith Chart (cases X and Y).

This conclusion is also consistent with models that start from a formulation based on patch antennas where it is recommended that broadband patches be given convex rounded profiles (Kumar and Ray, 2003). However, this new approach provided here also allows one to consider rounded or bevelled profiles as the synthesis of transmission lines having a continuously variable characteristic impedance. This technique is used in radio frequency for broadband matching. The characteristic impedance varies monotonically with differential increments for contiguous sections of line, between the input and output impedances. Thus, a knowledge of the aforementioned profiles allows us to systematically obtain UWB monopoles with maximised bandwidth.

6.2. Case Study: Maximising AMP Impedance Bandwidth

The aim is to design a monopole antenna with the greatest possible bandwidth. For simplicity, a PMA is chosen. As a hypothesis for this problem, the lower operating frequency limit is set at 1.4 GHz. To focus on one of the techniques in the previous section, this example will be based on using a bevelled lower edge. This is representative of other (rounded) profiles and is chosen because it is easier to set the parameters. The size of the ground plane is once more fixed at 260 mm × 260 mm.

As defined in the previous chapter, the frequency differences between the resonances added when steps or notches are made in the profile are inversely proportional to the width W of the PMA. Once the width W is fixed, (e.g., $W = 30$ mm), the spacing of the added resonances is fixed (5 GHz, given by the condition $W = \lambda/2$). The number is related to the number of steps. If the profile is bevelled or rounded, one assumes that it is theoretically infinite (an infinite number of steps). The difficulty lies in adjusting the spacing of the frequencies of these resonances through the W parameter to avoid a mismatch between them.

6.2.1. *Lower frequency limit of the band: Initial L, W and p parameters*

The height above the ground plane is mainly defined by the lower frequency according to the equations mentioned in the previous chapter. For a frequency of 1.4 GHz, a height $L = 45$ mm above the ground plane is defined.

As a general consideration, to design a broadband antenna, the half-width W must be increased from the linear monopole situation. This increase must be limited if other factors are taken into account, such as the omnidirectionality of the radiation pattern which is lost due to the resulting radial asymmetry. To fulfil both requirements, a compromise must necessarily be reached. The minimum width must therefore be chosen for which, when this bevel cut is made, the different resonances introduced satisfy the condition VSWR < 2. The greater the monopole W, the more likely it is that this condition will be satisfied as the same number of resonances will be introduced at closer frequencies.

The starting point is a relatively small half-width W (12 mm) and a starting value of parameter $p = 3$ mm. Both these values are the result of the optimisation of a rectangular PMA carried out in Sec. 4.4.7, with $L = 45$ mm (Fig. 6.9(b) (black)). Given this situation, the aim is that the following two conditions are fulfilled at the end of the matching process:

- The impedance plot is radially centred in the chart (it is currently displaced to the right, although the main lobe is centred).
- The area occupied by the impedance plot on the chart is the minimum possible (i.e., the added resonances overlap).

The influence of the W, p and l_y parameters on achieving this objective is shown below.

6.2.2. *Adjusting the width*

Starting from the initial situation $L = 45$ mm, $W = 12$ mm, $p = 3$ mm, when the transmission line is lengthened ($W = 28$ mm), the plot turns in a clockwise direction. This change does not have a negative effect on the initial condition for the lower operating frequency since this value is reduced.

Figure 6.9 shows points A and B, corresponding to $W = \lambda/2$, before and after the widening (12.5 GHz and 5.35 GHz, respectively). As one would expect, when the bevel cut is made, the resonances are accentuated at these points.

6.2.3. *Changing the height p over the ground plane*

The effect of making adjustments to the height p over the ground plane for the two previous values of W will now be considered. As explained in Chap. 4, when p is decreased, the characteristic impedance is decreased and the impedance plots spreads out over the chart (Fig. 6.10). The first condition for matching then is fulfilled (it becomes centred in the Smith Chart). The second (overlapping of new resonances) must now be satisfied, without significantly altering the first.

6.2.4. *Implementing a bevelled cut*

The effect of making a bevel-shaped cut is now studied (Fig. 6.4(b)). Its effect on the Smith Chart is shown in Fig. 6.11. Taking a half-width W of 12 mm, lobes

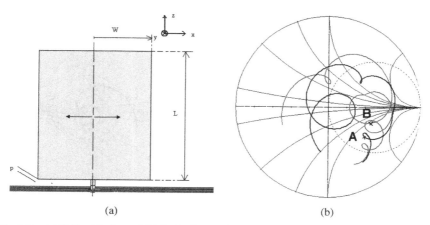

Fig. 6.9. (a) PMA widening and (b) Smith Chart (1–20 GHz). $L = 45$ mm, $p = 3$ mm, $W = 12$ mm (black) and $W = 28$ mm (grey). Points A and B at $W = \lambda/2$.

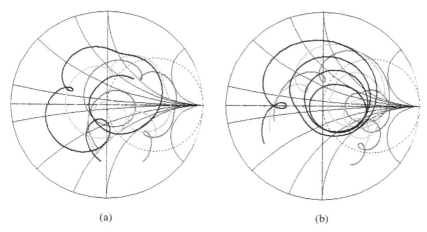

Fig. 6.10. Effect on the Smith Chart (1–20 GHz) of changing the distance p over the ground plane for a rectangular PMA. $L = 45$ mm, $p = 3$ mm (grey), 1.75 mm (light grey), 0.5 mm (black). (a) $W = 12$ mm, (b) $W = 28$ mm.

would be produced successively at 12.5 GHz, 25 GHz and so on, which is not ideal in this case because the frequencies are very widely spaced (Fig. 6.11(a)). One can observe that these local resonances also cause the whole plot of the impedance to be folded back on itself. This is due to the creation of the new localised lobes and their repercussions on adjacent frequencies. The bevelled shape also causes a third

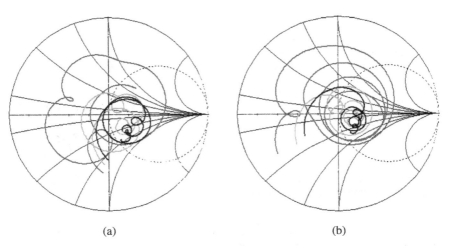

(a)   (b)

Fig. 6.11.   (a) Smith Chart (1–20 GHz). $W = 12$ mm, $L = 45$ mm, $p = 0.5$ mm; $ly = 0$ mm (grey), 4.75 mm (light grey), 9.5 mm (black). (b) Smith Chart (1–20 GHz). $W = 28$ mm, $L = 45$ mm, $p = 0.5$ mm; $ly = 0$ mm (grey), 7 mm (light grey), 14 mm (black).

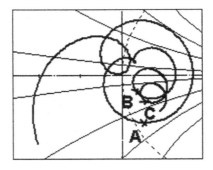

Fig. 6.12.   Detail of the impedance plot for the selected PMA: $p = 0.5$, $W = 28$ mm, $ly = 7$ mm.

undesirable effect of rotating or turning the plot because the path travelled by the current is lengthened.

Figure 6.11(b) shows that when $W = 28$ mm, the position of the added lobes is more advantageous.

The Smith Chart for the case where $W = 28$ mm and $ly = 7$ mm is presented in Fig. 6.12 amplified and shows the lobes corresponding to the frequencies at which $W = \lambda/2$ (5.35 GHz, point A), $W = \lambda$ (10.7 GHz, point B) and $W = 3\lambda/2$ (16 GHz, point C). This shows how lobes overlap each other, sometimes when the lobe is not completely closed (e.g., point A).

6.2.5. *Changing the profile close to the feed*

The outline shape of the disc at the lower part of a cylindrical monopole represents a capacitance in parallel with the antenna (Brown and Woodward, 1945). Changes in this capacitance between the base of the antenna and the ground plane have a considerable effect on the matching. This effect can be controlled by changing the parameter c in Fig. 6.13.

The result on the admittance Smith Chart of putting a capacitance in parallel with an arbitrary load is shown in Fig. 6.14(a). In this chart, we can see that points

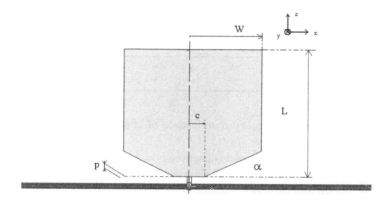

Fig. 6.13.   PMA-H and its parameters.

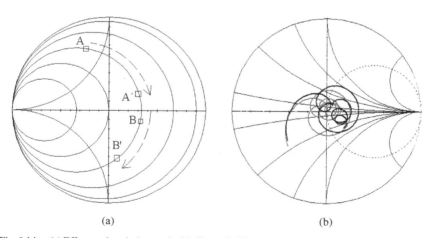

(a)   (b)

Fig. 6.14.   (a) Effect on the admittance Smith Chart of adding a capacitance in parallel with an arbitrary load, (b) Smith Chart (1–20 GHz). $W = 28$ mm, $L = 45$ mm, $p = 0.5$ mm; $ly = 7$ mm when $c = 0$ mm (black) or $c = 3$ mm (grey).

Table 6.2.   Summary of the changes and their effects on the impedance plot in the Smith Chart.

Change	Effect
W increased	Clockwise rotation with the centre on the right hand side of the Smith Chart.
p reduced	Plot spreads out and moves to the left.
ly — bevel cut	Resonances are added, the plot occupies a smaller area and rotates clockwise.
c — widening of the feed area	Plot rotates clockwise, with the centre on the left hand side of the Smith Chart.

Table 6.3.   PMA-H parameters (in mm and °).

L	W	α	p	c
45	28	16°	1	3

A and B turn clockwise along a circumference of constant admittance providing A′ and B′.

The effect on the chosen antenna of setting $c = 3$ mm is shown in Fig. 6.14(b). Here the same tendency is observed. Through this technique, an impedance plot that is more centred in the chart is achieved.

Table 6.2 offers a summary of the relationships between changes in the antenna's geometrical parameters and their repercussions in the Smith Chart. This may give rise to a systematic iterative process that ensures that bandwidth is maximised (Valderas *et al.*, 2006a).

The resulting antenna is a planar hexagonal antenna (PMA-H). The final design parameters, in accordance with Fig 6.13, are shown in Table 6.3.

These values result in a PMA that fulfils the initial specifications.

6.2.6. *Simulation and measurements*

6.2.6.1. *Impedance bandwidth*

The connector used in measuring the prototype is a 2.4 mm PC type with an operating range of 50 Ω up to 50 GHz. This is inserted through a hole in the central part of the ground plane. The supporting foot for the monopole, of length p and width 1.2 mm, forms part of the monopole. The groundplane is a square, 260 mm × 260 mm, so it does not negatively affect the bandwidth.

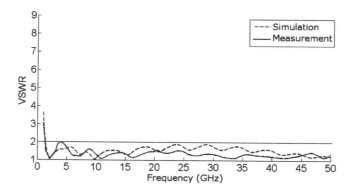

Fig. 6.15.   VSWR versus frequency for the PMA-H.

The comparison between the simulated and measured values for the VSWR is shown in Fig. 6.15.

The bandwidth achieved for VSWR < 2 is from 1.33 GHz to over 50 GHz. This gives a ratio of over 37:1 (Valderas *et al.*, 2006a). The upper limit of the band is determined by the limitations of the measuring equipment.

6.2.6.2. *Radiation patterns at representative frequencies*

The measured radiation patterns are shown for representative frequencies of the UWB bandwidth defined by the FCC (3.1 GHz and 10.6 GHz) in the three main cuts of the PMA-H, in Figs. 6.16 and 6.17, respectively. The corresponding axes are those indicated in Fig. 6.13.

The general shape is in line with the comments already made about the patterns for the PMA-S at lower frequencies.

If the separation $2W$ between the vertical edges is made to be close to $\lambda/2$ (0.58 λ in Fig. 6.16(c) for 3.1 GHz), a destructive interference is produced in the plane of the PMA in direction $\Phi = 0°$ or $\Phi = 270°$. In Fig. 6.17(c), for 10.6 GHz, this distance is 2λ; the interference is therefore constructive and the pattern tends to become oval along that axis. The local nulls that may occur in the H-plane between one frequency and another can be prevented by arranging crossed monopoles having revolution symmetry, as will be discussed in Chap. 8.

The crossed component of the fields is caused by the horizontal components of the current. This is why there is a null at $\Phi = 0°$ in the H-plane. As it is a symmetrical antenna, with horizontal currents going in opposite directions from the feed, there are also nulls for $\Phi = 90°$ and $\Phi = 270°$ in the H-plane. These conditions do not occur when the currents are no longer coplanar.

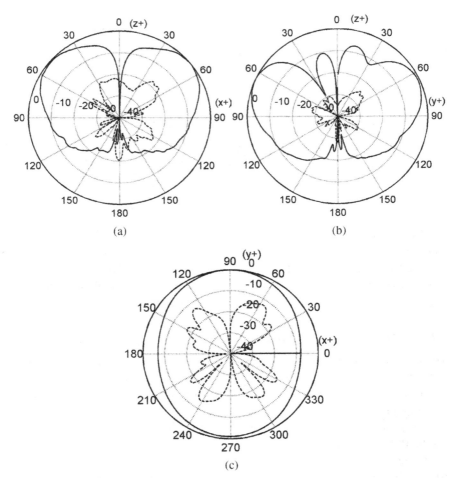

Fig. 6.16. (a) Measured radiation pattern for the PMA-H at 3.1 GHz in the $\Phi = 0°$ plane, (b) in the $\Phi = 90°$ plane and (c) in the $\theta = 90°$ plane. Co (-) and cross (--) polarisation.

The next section studies the repercussions that such a considerable bandwidth has on spectral efficiency, in comparison to the antennas previously designed.

6.3. Discussion of Spectral Efficiency in Broadband Antennas

Given that a range of different PMAs of differing bandwidths have been studied in this chapter and the preceding ones, this is the moment to analyse how the various UWB antennas use the available spectrum for radiating.

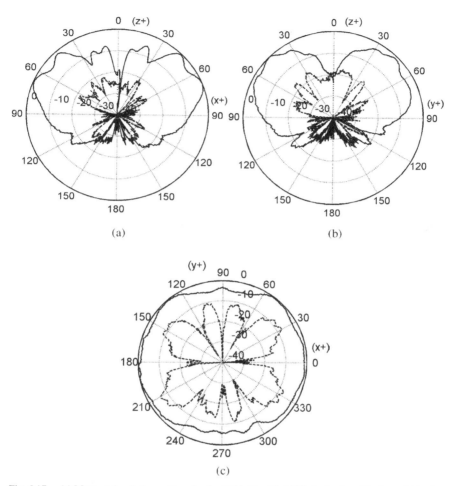

Fig. 6.17. (a) Measured radiation pattern for the PMA-H at 10.6 GHz in the $\Phi = 0°$ plane, (b) in the $\Phi = 90°$ plane and (c) in the $\theta = 90°$ plane. Co (-) and cross (- -) polarisation.

In Chap. 2, the spectral efficiency η_{rad} of an antenna was defined using Eq. (6.4).

$$\eta_{rad}(\%) = \frac{\int_0^\infty Pt(\omega)(1 - |\Gamma(\omega)|^2)d\omega}{\int_0^\infty Pt(\omega)d\omega} \times 100\%, \qquad (6.4)$$

where Pt is the power at the input to the transmitting antenna and Γ is the reflection coefficient normalised to the value of Zo for the study.

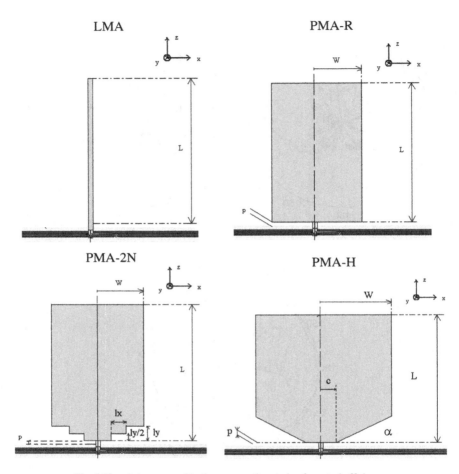

Fig. 6.18.   Antennas used in the comparative study of spectral efficiency.

Figure 6.18 shows the four antennas that are being compared: from the linear monopole (LMA) that is used as a reference and which is widened to become the rectangular planar monopole (PMA-R), to two known prototypes.

Table 6.4 shows the parameters for the four antennas.

Figure 6.19 shows the values of VSWR measured between 10 MHz and 50 GHz for the two reference antennas.

Figures 6.20 to 6.23 show the efficiency values for the four antennas as a function of σ of the pulse (from 10 to 110 ps), as defined in Eqs. (2.3) and (2.4). Each figure corresponds to a specific value of n.

Table 6.4. Parameters for the antennas in Fig. 6.18 (in mm and °).

	L	W	p	c	α	lx	ly
LMA	45	0.8	—	—	—	—	—
PMA-R	45	14	1.6	—	—	—	—
PMA-2N	45	14	1	—	—	4.7	5
PMA-H	45	28	1	3	16°	—	—

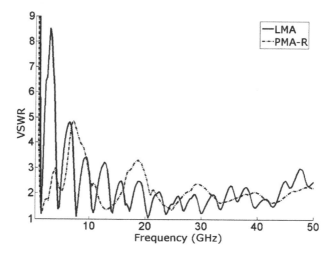

Fig. 6.19. VSWR versus frequency for the LMA and PMA-R.

In general it is seen that, as the order n of the derivative increases, the efficiency improves. This is partly conditioned by the lower limit of the antennas' bands, which can be considered to be fixed (typically 1.3 GHz). If the pulse is wider (σ is low) or is centred at higher frequencies ($n > 1$), this threshold is overcome and the efficiency tends to increase.

The antenna with maximised bandwidth (PMA-H) gives a higher value for the efficiency, independent of the pulse. It is also the most robust (over 90% for all Rayleigh pulses). When the emission limits are satisfied by the shape of the pulse used, this antenna always guarantees the maximum possible efficiency.[c] Of course, in this case only PMAs are being examined. These high efficiencies may also be

---

[c]As, for example with $n = 6$ and 76 ps $< \sigma <$ 106 ps for the mask established for UWB by the FCC (Chen *et al.*, 2004).

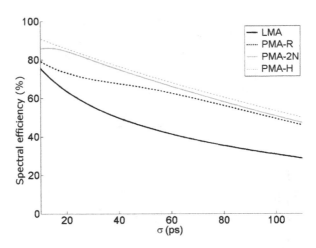

Fig. 6.20.   Spectral efficiency versus σ of the antennas considered for a Gaussian pulse ($n = 0$).

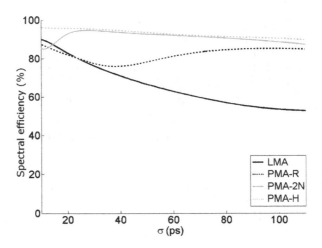

Fig. 6.21.   Spectral efficiency versus σ of the antennas considered for a Rayleigh pulse ($n = 1$).

accompanied by stable transfer functions and high levels of radiation uniformity when the antenna is given the appropriate shape and same bandwidth.

As expected, the LMA, having the narrowest band, tends to have the lowest values of efficiency. However, depending on the pulse used, the efficiency of the LMA can exceed that of a PMA-R.

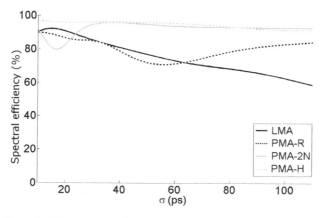

Fig. 6.22.   Spectral efficiency versus σ for the antennas considered for a Rayleigh pulse ($n = 3$).

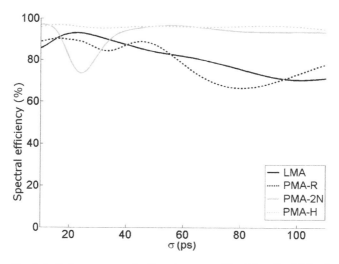

Fig. 6.23.   Spectral efficiency versus σ for the antennas considered for a Rayleigh pulse ($n = 7$).

The PMA-2N allows for greater efficiency than the PMA-R. However, for this to be the case one would have to take a pulse with a bandwidth centred on the impedance bandwidth (n > 0, σ > 40 ps). As the PMA-2N can function as a filter at certain frequencies, it is not surprising that its efficiency may be lower for increasing values of n: when $n = 7$, the maximum of the pulse comes into the filtered zone.

Chapter 7

UWB Folded Monopole Antennas

Daniel Valderas and Juan I. Sancho
CEIT and Tecnun, University of Navarra

This chapter covers the design of folded monopole antennas (FMAs) for UWB applications, starting from planar antennas and folding them along axes that are perpendicular to the ground plane. Folding a rectangular monopole changes the impedance Zl that is to be matched. Implementing notches or bevel-shaped cuts in this folded rectangular antenna will have the same consequences in terms of the forms of the Smith Chart plots as in the preceding cases. The techniques seen in previous chapters can therefore be applied, although the geometry of these antennas is not planar. Additionally, due to their geometry, they offer different radiation properties, e.g., improved omnidirectionality at high frequencies or improved directivity compared to planar monopole antennas.

7.1. Introduction

As the PMAs discussed in previous chapters lack radial symmetry, omni-directionality is also lost at higher frequencies. This originates from the concentration of radiating currents in the vertical edges and the difference in path distances travelled by the fields radiated by these currents. If it is assumed that these edges are strictly the ones that radiate, the interference will be destructive in the direction that is coplanar with the PMA at the frequency at which the width, $2W$, has the value $\lambda/2$. This becomes constructive when $2W$ equals λ. Meanwhile, the interaction is always constructive in the direction perpendicular to the monopole. As a result, omnidirectionality is lost in the H-plane due to the maximum migrating with frequency. We will call the frequency range in which this change occurs ($\lambda/2$ to λ), the **critical range for omnidirectionality** (CRO).

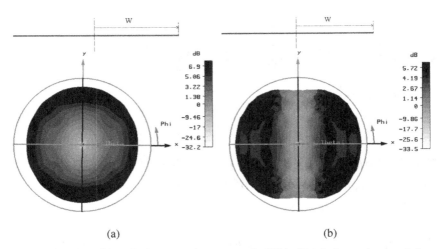

(a)   (b)

Fig. 7.1.   Top view of the radiation pattern for a rectangular PMA of height L over the ground plane. $L = 45$ mm, $W = 29$ mm, height of feed is 2.7 mm with a ground plane of 260 mm × 260 mm at a frequency such that (a) $2W = \lambda/2$ and (b) $2W = \lambda$ (5.17 GHz and 10.34 GHz, respectively).

Figure 7.1 shows the top view of the radiation pattern in which this phenomenon is observed, at frequencies for which $2W$ has the value $\lambda/2$ (on the left) and λ (on the right). The close link between the symmetry of the antenna and that of the radiation pattern can also be seen: in both cases this is with respect to the $\Phi = 0°$ and $\Phi = 90°$ axes.

There are various solutions to the excessive degradation of the radiation pattern's stability over frequency. One of these is to fold the antenna along vertical axes. This brings the edges of the antenna together and as a result, the frequencies at which the condition $2W$ equals $\lambda/2$ are higher (Fig. 7.2) (Wong et al., 2005b). In this case, the impedance bandwidth for VSWR < 2 and the omnidirectionality of

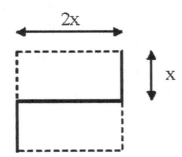

Fig. 7.2.   Top view of a monopole folded along vertical axes.

the radiation pattern, measured by the maximum variation in gain in a horizontal plane being less than 3 dB, have the same upper frequency limit (6.6 GHz). The lower operating limit is set by the impedance bandwidth (2.18 GHz). However, it is clear that this leads to bulkier antennas. Once more, a compromise solution must be reached.

An FMA presents some other advantages, such as reducing the transverse dimension $2x$ if this is required by the design specifications (Fig. 7.2). In the same way, directional antennas can be designed using folding. Furthermore, increasing the density of the folded configuration slightly reduces the height above the ground plane for the same lower frequency bandwidth limit (Wong *et al.*, 2005b).

Before tackling the design, it is necessary to analyse how currents are distributed in a FMA to evaluate whether the techniques seen previously based on TLM are still valid.

7.2. Current Distribution in an Antenna Folded along Vertical Axes

Figure 7.3 shows the distribution of horizontal currents in a rectangular PMA and in a configuration folded accordion-style at 1, 2, 5 and 10 GHz. When the **accordion antenna** (AA) is unfolded, it has the same W parameter as the PMA, so that

$$W = a + b + 2d, \tag{7.1}$$

where a, b and d are the fold dimensions as shown in Fig. 7.3. Its height L and separation from the ground plane p are likewise the same. Although the current distributions change, what is important for applying the model is to check that the working hypothesis seen in Chap. 4 (Sec. 4.2.1) remains true.

The horizontal currents in a PMA are coplanar. When folds are made in the antenna this condition is lost and the effect introduced by greater mutual coupling must be taken into account. This translates into different current distributions in the two faces of the monopole, but the horizontal current is preferentially concentrated in the lower part in both cases.

In general terms the vertical currents, shown in Fig. 7.4, are considerably smaller than the horizontal currents in the lower edge of the FMA. They may be greater at the feed point due to the condition it imposes, but this is the only area where this may happen.

7.3. TLM Applied to an FMA

As mentioned above, folding a PMA increases the mutual coupling between the currents. From the point of view of TLM, this translates into a modification to the

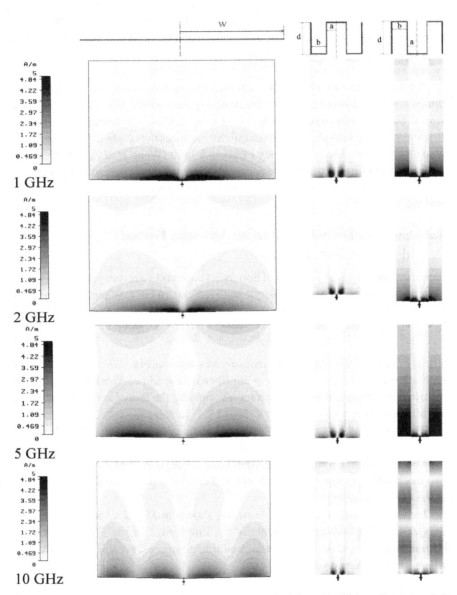

Fig. 7.3.   Horizontal currents in a PMA (left) and in AA on both faces (right). $L = 45$ mm, $p = 2.5$ mm, $W = 28$ mm, $a = 3$ mm, $b = 5$ mm, $d = 10$ mm.

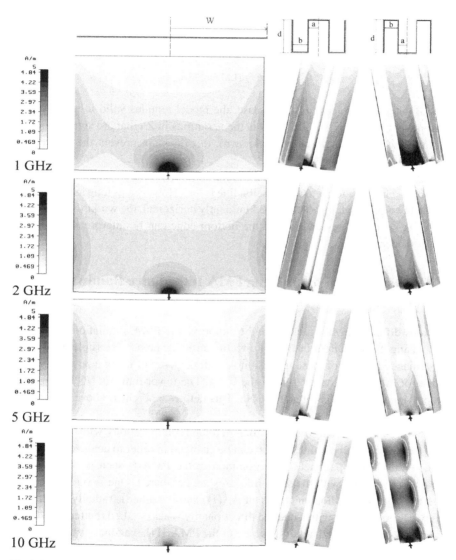

Fig. 7.4.   Vertical current distribution in a PMA (left) and in AA (right). $L = 45$ mm, $p = 2.5$ mm, $W = 28$ mm, $a = 3$ mm, $b = 5$ mm, $d = 10$ mm.

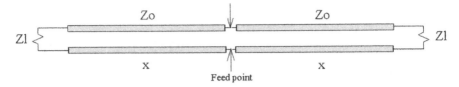

Fig. 7.5.   TLM for FMAs.

initial values of *Zl* and *Zo*. In any case, the model remains valid in terms of the parameters shown in Fig. 7.5. As it is the variations in *Zo* that are significant and not the initial states (which are unknown), this does not prevent the use of this model.

One could argue that, as the horizontal current is more spread out over the antenna, the character of a transmission line is lost. However, as long as the current in the **lower edge** of the FMA is predominantly horizontal, the working hypothesis remains valid and any changes to the bottom edge can be interpreted using the TLM model.

7.4. Case Study: Maximising FMA Impedance Bandwidth

7.4.1. *Design*

Various different FMAs are required to improve the PMA's radiation pattern by increasing its omnidirectionality or directionality; the greatest possible bandwidth is demanded for all of these. An example will be given of using a bevelled lower edge for different ways of folding the PMA. The lower limit for VSWR < 2, as in previous chapters, is set at 1.4 GHz. This defines a height *L* above the ground plane of 45 mm.

Figure 7.6 shows a top view of different folded configurations, with the fold axes and the geometrical parameters that can be changed in order to achieve maximum bandwidth. These parameters are common to the PMA-H seen in the previous chapter. Two types with the same fold axes are compared. One is symmetrical in the YZ plane (accordion antenna 1 or AA(1)) and the other is radially symmetric (AA(2)). An antenna with a marked directionality is also studied (butterfly antenna or BA). When unfolded, all of these are of the PMA-H (hexagonal) type (Valderas *et al.*, 2006b) as illustrated by their front view in the same figure.

To obtain the *W*, *p*, *α* and *c* parameters, the same criteria described in Chap. 6 for obtaining a PMA-H with maximum bandwidth are applied. The impedance plots for the three antennas, together with that of the PMA-H before (black) and after (grey) applying the methodology described in that chapter are shown in Fig. 7.7.

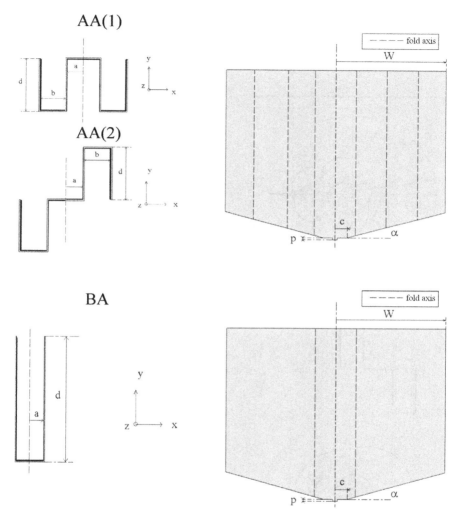

Fig. 7.6. Two variants of the accordion antenna AA(1) and AA(2) and the butterfly antenna BA with their respective fold axes and TLM parameters for matching. (©2006 IEEE.)

The initial situation to be matched is obtained by folding a rectangular PMA with the same width W as the PMA-H, as this PMA-H has a maximised bandwidth which is also aimed for in the folded configurations. The difference between the initial black plots in Fig. 7.7 originates exclusively from the different electromagnetic coupling between the currents, which in turn arises from the different ways in which the antennas are folded.

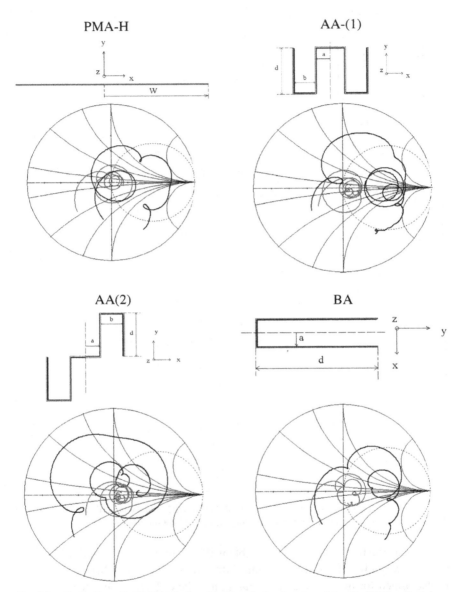

Fig. 7.7.   Smith Chart (1–20 GHz) before (black) and after (grey) applying the TLM methodology of Chap. 6 to planar and folded configurations. In the initial situations, $L = 45\,mm$, $p = 3\,mm$ and $\alpha = 0°$. For the AA, $a + b + 2d = W = 28$ mm. For the BA, $a + d = 28$ mm. (©2006 IEEE.)

Table 7.1.　Parameters of the FMAs (in mm and °).

	PMA-H	AA(1) and (2)	BA
L	45	45	45
W	28	38	33
p	1	1	1
α	16°	20°	12°
c	3	2	0
a	—	8	8
b	—	10	—
d	—	10	25

Table 7.1 gives the final parameters obtained by applying the method. Using these parameters, the matching obtained is such that VSWR < 2 over a maximised bandwidth, according to the measurements described in the following section.

This opens up the possibility using this method to design, without loss of bandwidth, a broad range of FMAs having diverse radiation patterns depending on how the antenna is folded. The upper limit of the band can also be controlled by using steps instead of bevels.

7.4.2. *Simulation and measurements*

7.4.2.1. *Impedance bandwidth*

Figure 7.8 shows the VSWR for the FMAs and the PMA-H.

For an initial specification of 1.4 GHz, the value of the bottom limit of the band where VSWR < 2 is 1.33 GHz for the PMA-H and slightly lower for the other designs: 1.3 GHz for the BA and 1.24 GHz for the AAs. This limit drops as *W* increases (width of the unfolded antenna), as one would expect. The upper limit exceeds the 50 GHz threshold. The typical bandwidth is greater than 38:1 (Valderas *et al.*, 2006b).

7.4.2.2. *Radiation patterns at representative frequencies*

Figures 7.9 to 7.14 show the radiation patterns for the FMAs at frequencies of 3.1 GHz and 10.6 GHz, which were chosen as representative of the bandwidth defined by the FCC. The axes are those in Fig 7.7.

In FMAs, cross polarisation increases in those planes for which symmetry is lost. This is not negligible in the $\Phi = 0°$ section for the BA and AA(1), but it remains at a low value for the $\Phi = 90°$ plane of symmetry. In the case of the

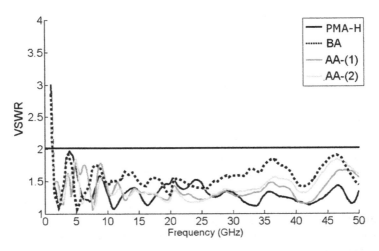

Fig. 7.8.   VSWR for the FMAs in Fig. 7.7 with the dimensions given in Table 7.1.

AA(2), for which symmetry has been lost in both planes, this may be of the same order as the directly polarised component.

As seen in the sections in the H-plane, the shape of the pattern is closely related to the way the antenna is folded.

These diagrams are a first approximation for studying the radiation patterns in general terms. However, it will be necessary to turn to the parameters defined in Chap. 2 for a more global appreciation of how variable they are over frequency. Some are covered in following sections, in comparison with the PMA-H.

7.4.2.3. *Uniformity*

7.4.2.3.1. Uniformity depending on the critical range for omnidirectionality

When the antenna is small in electrical terms, omnidirectionality is guaranteed for frequencies below the CRO. As the frequency increases, there is a loss of uniformity. Figure 7.15 compares uniformity as defined in Chap. 2 over a frequency band that corresponds to a separation between vertical edges of between $\lambda/2$ (2.67 GHz) and λ (5.34 GHz) for the PMA-H.

Table 7.2 shows the frequency at which uniformity drops to 90% ($f_{U=90\%}$) for the PMA-H and the AA(2), in comparison with the frequency at which the separation between edges is $\lambda/2$ ($f_{\lambda/2}$). Approximately the same ratio is maintained in both cases. Additionally, in accordance with Fig. 7.15, one can see that maintaining radial symmetry is beneficial for increasing $f_{U=90\%}$ ($f_{U=90\%\ AA(2)} >$

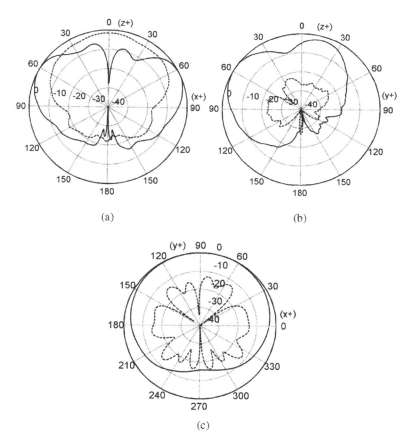

Fig. 7.9. (a) Measured radiation pattern for the BA at 3.1 GHz in the $\Phi = 0°$ plane, (b) in the $\Phi = 90°$ plane and (c) in the $\theta = 90°$ plane. Co (-) and cross (- -) polarisation. (©2006 IEEE.)

$f_{U=90\%\,AA(1)}$). However, increasing $f_{U=90\%}$ increases the frequency range over which uniformity is lost due to the migration of the maximum from $\Phi = 90°$ to $\Phi = 0°$.

The interval studied in Fig. 7.15 corresponds to Table 7.3 which reflects measured radiation patterns. As shown in the table, the first maximum persists in an intermediate situation (e.g., 4.4 GHz (0.82 λ)) along with the new one. Between the two, there is a dip that distorts the pattern. Then again, one can observe how the shape of the pattern and its nulls follow the symmetry set by the antenna. For example, in the case of antennas that are symmetrical about a plane (PMA-H, AA(1) and BA with respect to the $\Phi = 90°$ plane), the patterns and the nulls will follow the symmetry. Additionally, the nulls oscillate around relatively stable positions.

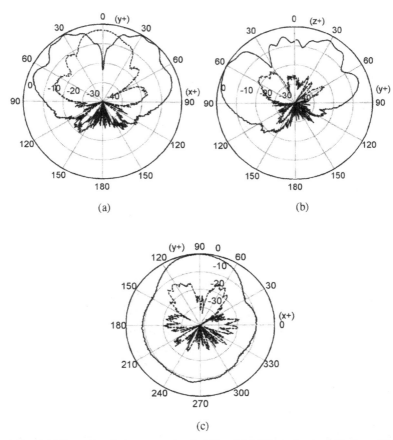

(a)   (b)

(c)

Fig. 7.10.   (a) Measured radiation pattern for the BA at 10.6 GHz in the $\Phi = 0°$ plane, (b) in the $\Phi = 90°$ plane and (c) in the $\theta = 90°$ plane. Co (-) and cross (--) polarisation. (©2006 IEEE.)

In the case of antennas that are radially symmetrical, but not symmetrical with respect to a plane (AA(2)), there is a rotation of the radiation nulls at the CRO. The measurements also show how the shape of this pattern in this plane, at the frequencies at which omnidirectionality is lost, is conditioned by how the antennas are folded.

The BA has two vertical radiating edges at 25 mm from the XZ plane (Fig. 7.16). The plane with sides L by $2a$ would act as a reflector at certain frequencies, increasing gain in the $\Phi = 90°$ direction.

The frequency at which a directional behaviour is obtained is estimated by Eq. (7.2), which accounts for the phase variation corresponding to the reflection in

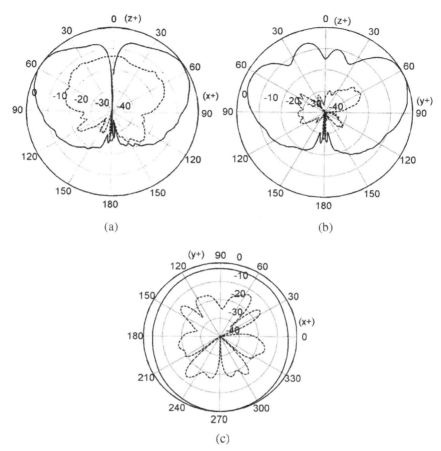

Fig. 7.11. (a) Measured radiation pattern for the AA(1) at 3.1 GHz in the $\Phi = 0°$ plane, (b) in the $\Phi = 90°$ plane and (c) in the $\theta = 90°$ plane (c). Co (-) and cross (- -) polarisation.

a metallic surface ($\lambda/2$). With $n = 1$, the frequency estimated in this case would be 3 GHz, from which frequency the uniformity falls below 60% (Fig. 7.15).

$$2d + \frac{\lambda}{2} = n\lambda. \tag{7.2}$$

7.4.2.3.2. Uniformity beyond the critical range for omnidirectionality

As the bandwidth at frequencies over the CRO has been optimised using TLM, it makes sense to study the effects of folding antennas on their omnidirectionality in that range. Fig. 7.17 shows uniformity up to 20 GHz.

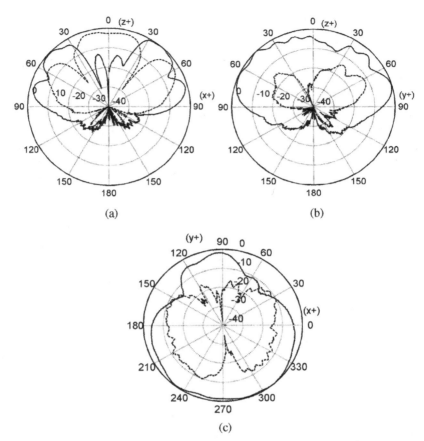

Fig. 7.12. (a) Measured radiation pattern for the AA(1) at 10.6 GHz in the $\Phi = 0°$ plane, (b) in the $\Phi = 90°$ plane and (c) in the $\theta = 90°$ plane. Co (-) and cross (--) polarisation.

One observes, for example, that once the CRO for each antenna has been exceeded, it is precisely the PMA-H that exhibits the best uniformity characteristics over a wider spectrum. However, when selecting the most appropriate antenna for the planned UWB operating range, one must also take other parameters into account.

Additionally, one can appreciate how Eq. (7.2), which defines successive directivity conditions for the BA, makes them predictable for different values of n. Thus, they will occur when d is equal to $3\lambda/4$ and $5\lambda/4$. In Fig 7.17, new dips in uniformity (peaks in directivity) are observed for the BA around these frequencies (9 GHz and 15 GHz, respectively).

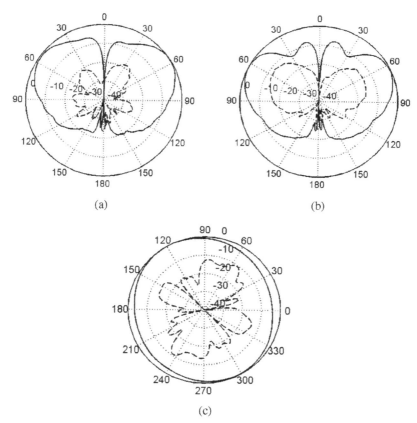

Fig. 7.13.   (a) Measured radiation pattern for the AA(2) at 3.1 GHz in the $\Phi = 0°$ plane, (b) in the $\Phi = 90°$ plane and (c) in the $\theta = 90°$ plane. Co (-) and cross (--) polarisation.

7.4.2.4. *Gain*

Figure 7.18 shows the absolute calculated gain, as defined by the IEEE, in dB for the H-plane of the FMAs considered. This is displayed over a wide spectrum in order to observe the variation in gain for these antennas. In sections $\Phi = 0°$ and $\Phi = 90°$, the gain is greater. Additionally, this parameter and those following serve to complete the significance of uniformity, since in each case the uniformity is normalised to a different value, depending on each antenna and frequency. Uniformity provides a study of the level of variability relative to the reference value set by the gain.

This shows that there is no geometry that increases gain stability in the H-plane and that all of them are similar over the range in question.

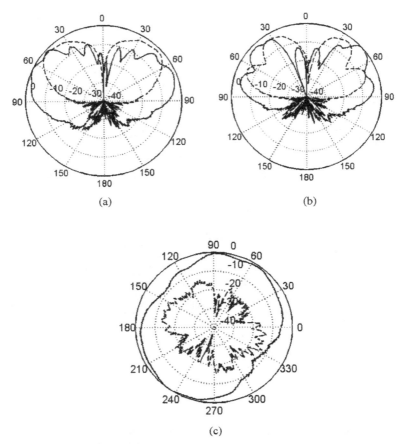

Fig. 7.14.   (a) Measured radiation pattern for the AA(2) at 10.6 GHz in the $\Phi = 0°$ plane, (b) in the $\Phi = 90°$ plane and (c) in the $\theta = 90°$ plane (c). Co (-) and cross (--) polarisation. (©2006 IEEE.)

7.4.2.5. *Transfer function*

Figure 7.19 shows the SATF magnitudes in the H-plane for the antennas being considered. For the purposes of comparison, the PMA-H is also included in the study. The radiated electrical field E_θ is calculated in transmission at 1100 mm under far-field conditions, as a response to a Gaussian pulse. As an average, stability depends both on frequency and on the type of antenna under consideration. There is therefore a range of different FMAs available, from which the most

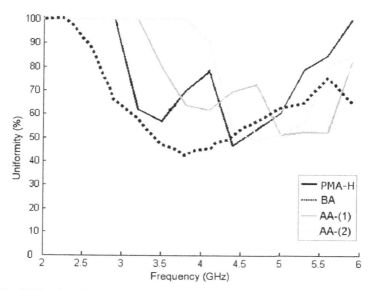

Fig. 7.15.   Uniformity of the θ component, both measured and simulated, in the H-plane for the PMA-H and the FMAs (2–6 GHz).

Table 7.2.   Comparison of $f_{\lambda/2}$ and $f_{U=90\%}$ calculated for the PMA-H and the AA(2).

	PMA-H	AA(2)
$f_{\lambda/2}$(GHz)	2.67	4.16
$f_{U=90\%}$ (GHz)	3	4.25

appropriate can be chosen for the operating range of present and future UWB technologies.

Similarly, it is clear that, in comparison with the SATF for the PMAs seen in Fig. 5.30, the dip at the frequency defined by the $W + L = \lambda$ condition is softened to a considerable extent by the fact that the new antennas are folded. Thus, more stable SATF profiles are obtained using this technique.

7.4.2.6. *Group delay*

This section studies SAGD stability, in order to check that these designs offer the most consistent behaviour possible over frequency.

Table 7.3.   Evolution of measured radiation patterns in the H-plane with frequency (GHz) for the PMA-H and the FMAs.

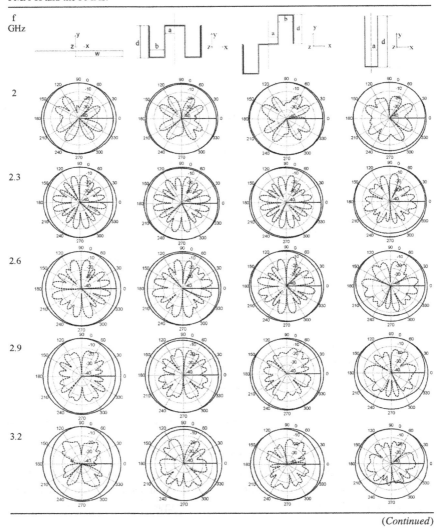

(Continued)

The average is calculated similarly to how it is calculated for the SATF over the same interval. Figure 7.20 shows that, independently of the type of FMA and due to the fact that they are monopoles with a common phase centre, SAGD stays stable over the spectrum, around a typical value. This characteristic is beneficial for UWB applications.

Table 7.3.   (*Continued*)

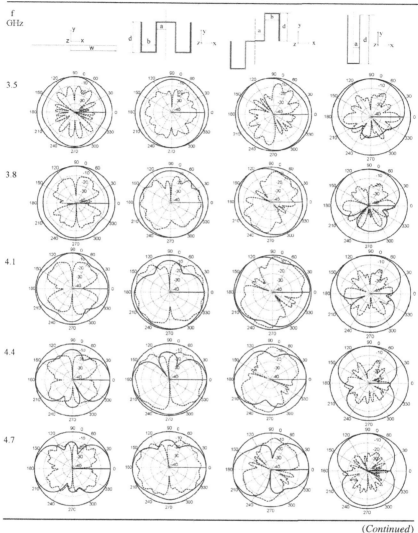

(*Continued*)

7.4.3. *Design options*

Taking into account that the FCC has defined a UWB transmission as one for which the fractional bandwidth is 20% or the occupied bandwidth is 500 MHz, respectively, the operating ranges are those selected by Table 7.4 in each case

Table 7.3.   (*Continued*)

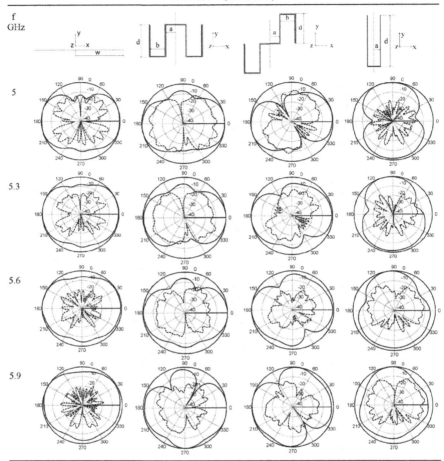

(Valderas *et al.*, 2006b). TLM offers a bandwidth from the FMA family in which to select the antenna offering the most stable parameters in terms of pulse: matching, uniformity, gain, SATF and SAGD in the H-plane. This allows the corresponding pulse in that range to be transmitted with maximum spectral efficiency and minimum distortion. It is also possible to obtain directional and omnidirectional antennas by setting the CRO in the appropriate range. For example, it is shown

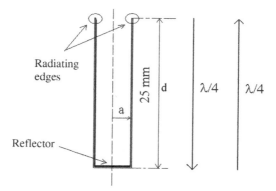

Fig. 7.16. BA studied as two radiating edges with a reflector and the case of constructive interference.

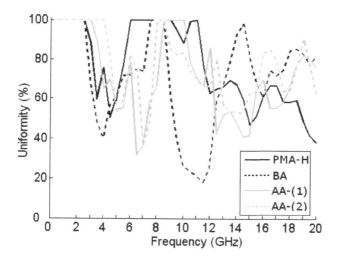

Fig. 7.17. Calculated uniformity of the θ component in the H-plane for the PMA-H and the FMAs.

that the AA(2) maintains better than 90% uniformity from the lower limit of the impedance bandwidth (1.24 GHz) up to 4.2 GHz, which gives the 3.4:1 ratio laid down by the FCC regulations on UWB. By suitably scaling the antennas, their operation can be customised, placing the optimal bandwidth wherever present or future pulsed transmissions may be defined.

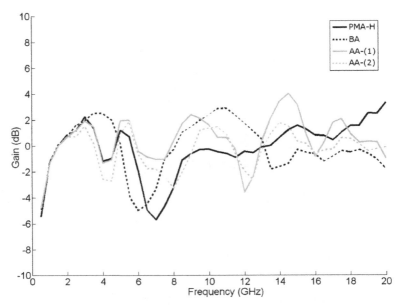

Fig. 7.18.   Calculated absolute gain in the H-plane for the PMA-H and the FMAs that have been designed.

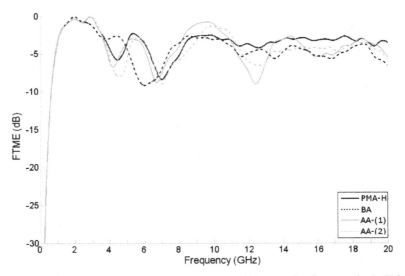

Fig. 7.19.   SATF (dB) of the θ component in the H-plane at 1100 mm against frequency for the PMA-H and the FMAs that have been designed.

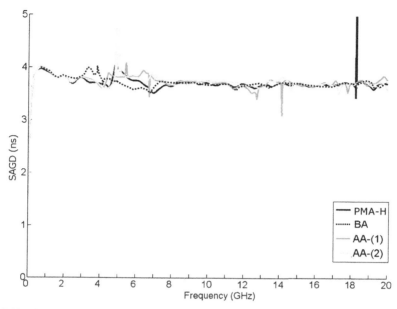

Fig. 7.20.   SAGD (ns) in the H-plane at 1100 mm against frequency for the PMA-H and the FMAs that have been designed.

Table 7.4.   Selection of bandwidths and antennas by stability of UWB operating parameters, taken from the total 38:1 VSWR < 2 bandwidth.

Bandwidth (GHz)	Antenna selected	VSWR < 2	Variation in H-plane gain (dB)	Uniform (%)	Variation in SATF (dB)	SAGD 1100 mm (ns)	Type of pattern
1.24–4.2	AA(2)	√	4.1	> 90%	6.9	3.7	Omnidir.
5.7–11.5	PMA-H	√	5.4	> 77%	5.7	3.7	Omnidir.
10–12	BA	√	1.1	< 26%	2.4	3.7	Directnl.
8–11	AA(1)	√	1.7	> 70%	3.3	3.7	Omnidir.
13.1–20	BA	√	1.9	> 70%[a]	2.8	3.7	Omnidir.

[a]Except for 15.4–16.4 where 60% < uniformity < 70%

In light of these considerations, the FMA design philosophy should be based on ensuring a good matching response through the use of TLM and selecting, from the potential range (38:1), the shape for which the other design parameters are best satisfied. Analogously, staircase profiles could be used instead of bevelled shapes.

Chapter 8

Revolution Monopole Antennas

Daniel Valderas and Juan I. Sancho

CEIT and Tecnun, University of Navarra

As considered in Chap. 4, one technique for improving the radiation pattern's omnidirectionality in the H-plane consists of setting two PMAs at right angles in the form of a cross. These are Revolution Monopole Antennas (RMAs) in which the antenna's radial symmetry is accentuated. This chapter covers the design of antennas with this topology using TLM.

After analysing the current distribution, different examples are designed, each giving improved omnidirectionality. The radiation in the H-plane and the corresponding transfer functions are then studied.

8.1. Introduction

One way to improve omnidirectionality in the H-plane is to give the antenna greater cylindrical symmetry by using several crossed PMAs (i.e., at angles to each other, while having the same axis of revolution) to produce a RMA. The example in Fig. 8.1 shows an RMA offering an orthogonal configuration.

The PMAs used in this configuration are of several types, including square monopoles with a semi-circular base and those using the bevelled edge technique (Anob *et al.*, 2001; Ammann and Cordoba, 2004b). U-shaped slots will also be implemented on these antennas to filter undesirable bands (Lee *et al.*, 2006).

It should be emphasised that the lower frequency bandwidth limit is basically maintained from the case of a single monopole to the crossed monopoles. The improvements in the pattern are considerable. However, when PMAs are crossed, the broadband-matching properties are sometimes lost (Anob *et al.*, 2001).

By combining FMAs and RMAs, new designs are produced, as shown in Fig. 8.2. This also has notches in the lower edge to increase the impedance bandwidth (6.45:1). This gives almost omnidirectional radiation patterns in the H-plane.

Fig. 8.1.   Top view of a RMA antenna made up of crossed PMAs.

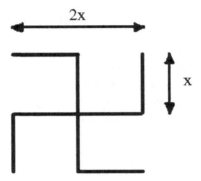

Fig. 8.2.   Top view of a PMA-based folded RMA (Wong *et al.*, 2005b).

Analogous results are obtained when three plates are set 120° apart instead of the four at 90° using the same notch technique to improve matching (Wong *et al.*, 2004b).

8.1.1. *Current distribution in an RMA*

Figures 8.3 and 8.4 show the current distribution in an RMA such as the one in Fig. 8.1, in comparison with that for a PMA. There is no appreciable change in the horizontal components. The apparent reduction in horizontal currents in the revolution monopole version is not, in fact, a reduction in total horizontal currents, as there is another plate connected in parallel (Fig. 8.3). It is clear that the horizontal current components are clearly concentrated in the lower edge and in that area they are greater than the vertical components (Fig. 8.4).

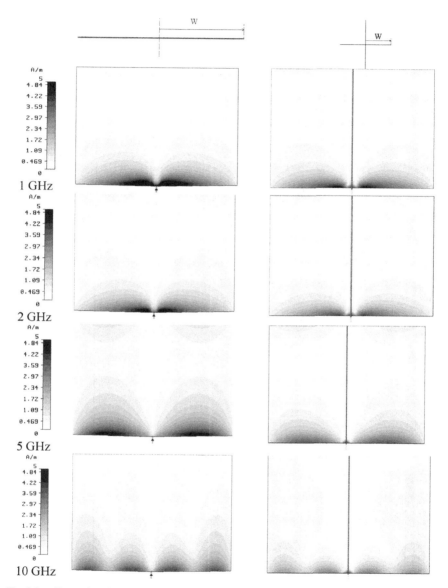

Fig. 8.3.   Comparison between horizontal current distribution in a PMA (left) and an RMA (right) at different frequencies. Height over ground plane $L = 45$ mm, distance to ground plane $p = 2.5$ mm, $W = 28$ mm.

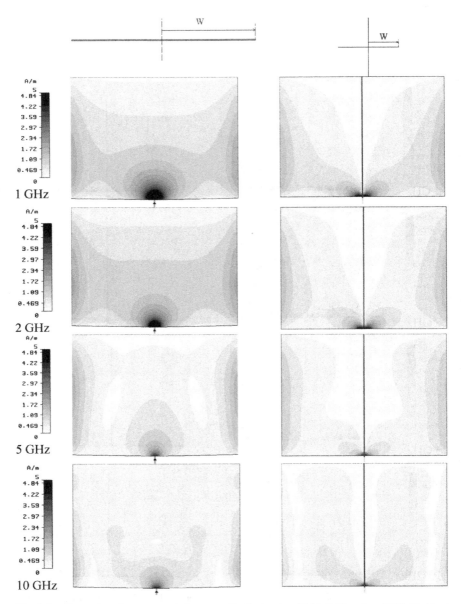

Fig. 8.4.   Comparison between vertical current distribution in a PMA (left) and an RMA (right) at different frequencies. Height over ground plane $L = 45$ mm, distance to ground plane $p = 2.5$ mm, $W = 28$ mm.

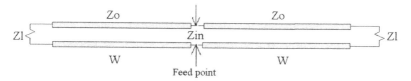

Fig. 8.5.   TLM analogy of a PMA.

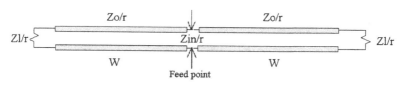

Fig. 8.6.   TLM analogy of an RMA with revolution factor r.

8.1.2. *TLM applied to an RMA*

The **repetition factor** r is defined as the number of times the monopole antenna is repeated at equidistant angles. The greater the omnidirectionality required, the higher this factor should be.

From the TLM point of view, an RMA with revolution factor r can be understood as parallel to that shown in Fig. 8.5.

If in this parallel, the Zo and Zl parameters are maintained, i.e., if the current distribution in each plate is the same before and after conversion into an RMA, the RMA case would be equivalent to simply dividing the Zo and Zl parameters by r, as shown in Fig. 8.6. The input impedance Z_{in} would also be divided by r.

Therefore, as r increases, the impedance plot would "spread out" over the Smith Chart towards areas of lower normalised impedance magnitude. In this case, any change to the lower profile would be made to each of the r monopoles making up the RMA. The extreme case would consist of a repetition factor r equal to infinity. In this case, the revolution antenna becomes a solid cylinder with a radius equal to the semiwidth W and a modified profile in the feed area.

8.1.3. *Case study: Maximising RMA impedance bandwidth*

8.1.3.1. *Design*

The aim is to design several different RMAs with the greatest possible bandwidth and no less than that given by the PMA on which they are based. An example will be given using a bevelled lower edge for different RMAs. The lower limit

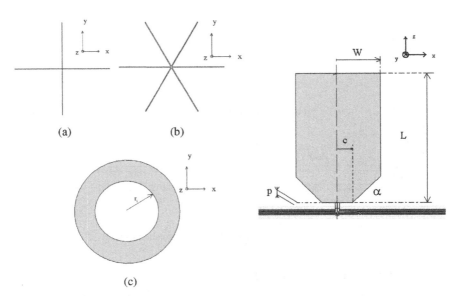

Fig. 8.7.   Left: RMAs having different values of r (a) RMA-(2) with $r = 2$; (b) RMA-(3) with $r = 3$; (c) RMA-(inf) with $r = \infty$. Right: One sheet of the RMAs with its respective TLM parameters for matching.

for VSWR < 2, as in previous chapters, is set at 1.4 GHz. This defines a height L above the ground plane of 45 mm. In this case, the versatility of the family of RMAs provides the desired degree of omnidirectionality: the greater r is, the greater the omnidirectionality. This improvement will be evaluated based on the statistical parameters considered in Chap. 2.

Figure 8.7 shows the top view of different configurations, together with the parameters used to achieve the maximum bandwidth, which are common to the PMA-H.

Any changes made to the W, p, α and c parameters in order to achieve the objective in each case are subject to the same criteria shown in Chap. 6 for the PMA-H, independently of the value of r. Figure 8.8 shows the impedance plots for the three antennas before and after matching together with that for the PMA-H. The PMA-H can be considered an RMA with $r = 1$. RMA-(inf) has $r = \infty$. One can see how, as r increases, the black trace corresponding to the initial situation (RMA made up of rectangular PMAs) spreads out due to the reduction in Zl as the number of transmission lines connected in parallel increases. This is in agreement with the model shown in Fig. 8.6.

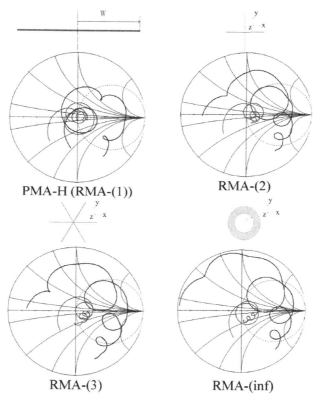

PMA-H (RMA-(1)) RMA-(2)

RMA-(3) RMA-(inf)

Fig. 8.8. Smith Chart (1–20 GHz) before (black) and after (grey) maximising the bandwidths of the PMA-H and RMAs with the initial conditions $L = 45$, $W = 28$ mm, $p = 3$ mm, $\alpha = 0°$.

Table 8.1. Parameters of the RMAs (in mm and °).

	RMA-(2)	RMA-(3)	RMA-(inf)
L	45	45	45
W	19	16	9.5
p	1	1	1
α	37°	45°	56°
c	0	0	0
r_i	—	—	5

Table 8.1 shows parameters W, p, α and c obtained for these configurations (Valderas *et al.*, 2007b). This shows clearly that as r increases, their values are changing in the same direction. On one hand, the antenna tends to become narrower as r increases, occupying a smaller volume. On the other, the angle α or bevel angle must be increased to compensate for the reduction in Zo caused by the increase in r (Fig. 8.6). Additionally, the W ratio for the RMA-(1) and RMA-(inf) antennas (28 mm and 9.5 mm, respectively) is $2.95 \approx \pi$, as has been considered in the concept of equivalent radius applied to PMAs (Kumar and Ray, 2003).

For the antennas shown, it is not necessary to make a fine adjustment using the parameter c. Additionally, as a consequence of the current distribution analysis referred to, the central part of the antennas which hardly conducts any current is not required. It is easiest to remove this part in RMA-(inf) antennas. This is the meaning of the parameter r_i, which allows the weight of the antenna to be reduced without negatively affecting the bandwidth.

This method therefore opens up the possibility of designing a very wide range of RMAs with increasing values of r without loss of bandwidth. This results in radiation patterns showing increasing omnidirectionality. The upper limit of the band can also be controlled by using notches instead of bevels.

8.1.3.2. *Simulation and measurements*

8.1.3.2.1. Impedance bandwidth

Figure 8.9 shows the VSWR for the antennas designed over a 260 mm × 260 mm ground plane that satisfy the initial specifications with the parameters shown in Table 8.1. As for the FMAs, the lower bandwidth limit drops slightly when a plate is used in a revolution antenna (typically 1.2 GHz, compared to 1.33 GHz). Thus, in the case of the RMA with $r > 1$, the bandwidth ratio increases to at least 41.7:1 (Valderas *et al.*, 2007b).

8.1.3.2.2. Uniformity

As the aim of the revolution configuration is to achieve omnidirectionality, H-plane sections will continue to be used to check that this is being achieved. Figures 8.10 and 8.11 show uniformity for two threshold values: 6 dB for operational utility analysis and 10 dB for locating nulls, respectively.

Both graphs show the improved uniformity in a higher bandwidth introduced by increasing r. This is due both to the reduction in W (19, 16 and 9.5 mm, respectively) and to the fact that the angular separation between adjacent plates

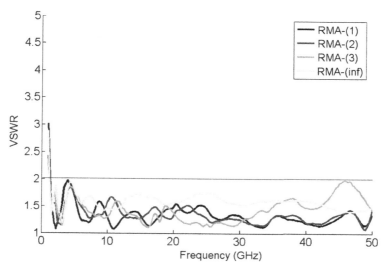

Fig. 8.9. VSWR for the RMAs in Fig. 8.7 having the parameters given in Table 8.1 and for the RMA-(1), which is equivalent to the PMA-H.

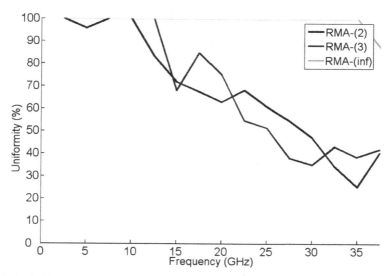

Fig. 8.10. Uniformity for the RMAs of the θ component in the H-plane measured in an anechoic chamber (2.6–37.5 GHz).Threshold value is 6 dB.

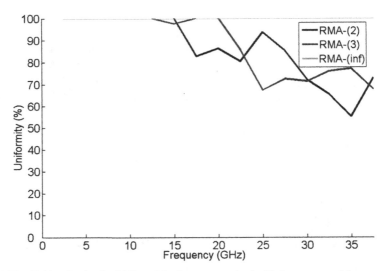

Fig. 8.11.   Uniformity for the RMAs of the θ component in the H-plane measured in an anechoic chamber (2.6–37.5 GHz). Threshold value is 10 dB.

of the antenna is reduced. Table 8.2 shows the corresponding radiation patterns measured at 2.5 GHz intervals between 2.6 and 37.5 GHz.

Taking a more general view, it is clear that when the distance between the two vertical edges of two adjacent plates tends towards the wavelength, the pattern in the H-plane tends to take the shape dictated by the antenna as the frequency is increased. This starts to occur around 12.5 GHz for the RMA-(2) and from 22.5 GHz for the RMA-(3). The general improvement in uniformity between both cases covers this range. The ratio of these two frequencies is 1.8. In a first approximation, the ratio is similar to that of the r parameter (3/2), taking into account that the W parameter is not the same as in the two cases. Once this threshold has been passed, when the patterns have taken on the antenna's shape, there is no reason why the uniformity should be better in one case than in the other. At the limit where r is ∞, the improvement theoretically takes place over a bandwidth that is also infinite.[a]

---

[a]Ideally, for the RMA-inf, U should be 100% at all frequencies. However, in Fig. 8.10, a reduction can be seen at higher frequencies. This is due partly to the diffraction at the edges of the squared ground plane, and partly to the fact that this antenna is heavier, so that it bends slightly when the z-axis is in a horizontal position when measurements are made.

Table 8.2.   Evolution of radiation patterns in the H-plane for RMA-(2), RMA-(3) and RMA-(inf).

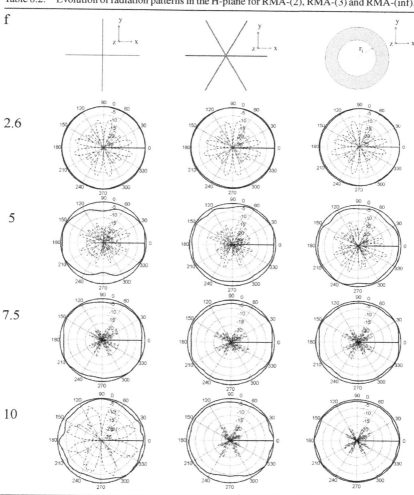

(*Continued*)

8.1.3.2.3. Gain stability

Figure 8.12 shows the absolute gain as defined by the IEEE for the θ component in the azimuth plane of the RMAs. One can see that the improvement in uniformity

Table 8.2.   (*Continued*)

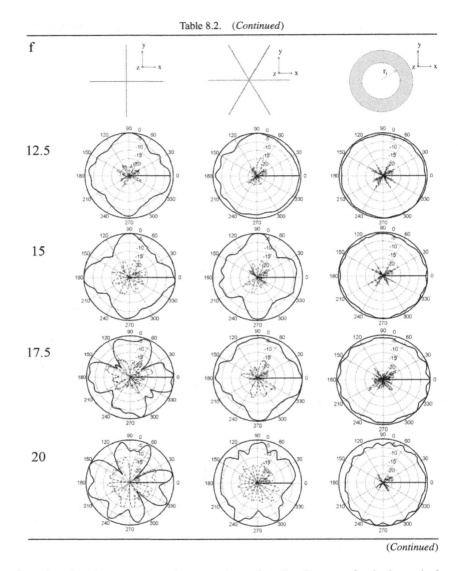

(*Continued*)

introduced as the *r* parameter increases is produced at the cost of reducing gain for the component of interest by between 2 and 3 dB over the whole range.

The figure also shows that the gain drops at around 5 GHz. As has already been emphasised, this is due to the monopole configuration. A cylindrical monopole with a negligible cross-section compared to the wavelength has a radiation null in the whole of the H-plane at the frequency at which the height is λ. If the

Table 8.2. (*Continued*)

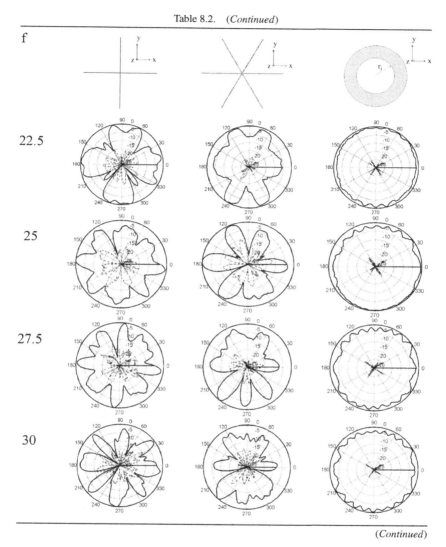

(*Continued*)

cross-section of the antennas were negligible, this would occur at 6.6 GHz, according to Eq. (5.11). Depending on the antenna's diameter, this would occur at different frequencies around 5 GHz.[b]

---

[b]As the RMAs under consideration have different radii, the dip at 5 GHz is more or less pronounced, depending on the antenna. One would expect the magnitude of the dip to be the same at adjacent frequencies. This cannot be seen in the graph, as the points have been taken every 2.5 GHz.

Table 8.2.   (*Continued*)

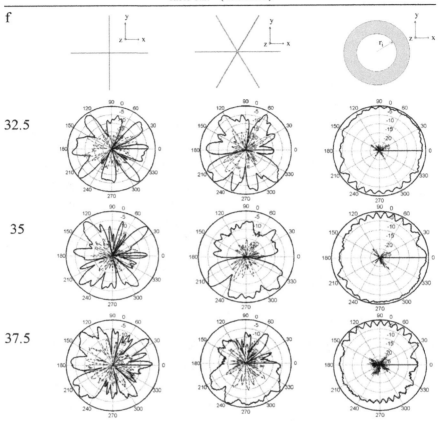

Once this critical point has been passed, the gain oscillates between ±2 dB for the RMA-(2) and RMA-(3) and between −1 dB and −4 dB for the RMA-(inf). A suitable design must take the *r* factor into account when optimising omnidirectionality and gain at the same time.

8.1.3.2.4. Transfer function stability

Figure 8.13 shows the magnitude of the values of SATF in the H-plane for the antennas under consideration. These were obtained in the same manner as those in the previous cases. The radiated electrical field E_θ is calculated in transmission at 1650 mm under far-field conditions, as a response to a Gaussian pulse. As the aim

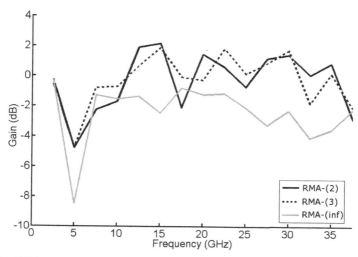

Fig. 8.12.   IEEE gain for the θ component versus frequency, measured in the H-plane for the three RMAs.

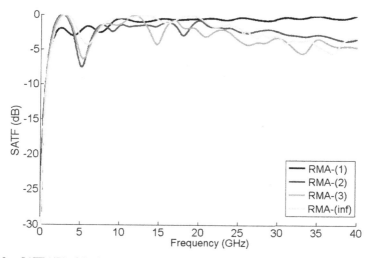

Fig. 8.13.   SATF (dB) of the θ component in the H-plane versus frequency at 1650 mm for the RMAs.

is to cover a wider spectrum (up to 40 GHz), this distance has been increased to give far-field conditions in all cases.

Apparently there is no close correlation between the stability of the function and the r factor. A more detailed observation of the phenomenon would be achieved if

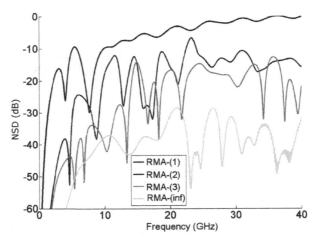

Fig. 8.14.   Normalised standard deviation (dB) of the θ component in the H-plane for the RMAs versus frequency at 1650 mm.

the transfer function were considered not only from the point of view of its average value in the H-plane (SATF), but also from the standard deviation of the transfer functions taken in different directions. Figure 8.14 shows the estimated value of the normalised standard deviation (NSD). This has been normalised with respect to the maximum value, not for each separate antenna (as in the case of Fig. 8.13), but for all the antennas. In this way, the variability with respect to one sole reference point can be compared. A normal distribution is assumed, with a 95% confidence interval for estimating parameters. This deviation with frequency clearly drops as r increases. The angular dependency typical of UWB-type transmissions, in the H-plane, can therefore be eliminated for antennas with a high r factor.

8.1.3.2.5. Group delay stability

Figure 8.15 shows the SAGD in the H-plane. The average is taken in the same way as for the SATF. Once more, the phase linearity of all the antennas, including the PMA with $r = 1$, around a typical value (5.5 ns) is clear.

8.1.3.3. *Design options and conclusions*

In short, it has been experimentally shown that the desired degree of uniformity for the RMAs in the H-plane is controlled by the r revolution factor. This degree of uniformity is achieved in the increase in the upper bandwidth limit with increasing r for a specific level of uniformity, e.g., 90%. The improvement in this aspect of the

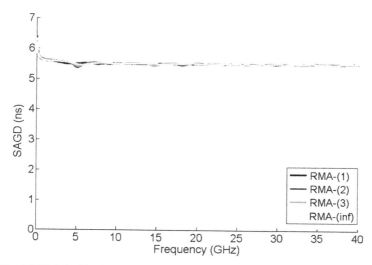

Fig. 8.15.   SAGD (ns) of the θ component in the H-plane versus frequency at 1650 mm for the RMAs.

RMA-(inf) has a negative impact on the antenna's gain in the H-plane, reducing it by about 3 dB over the whole range compared with the other prototypes. In this case, the SATF in the H-plane is not sufficient to analyse the improvement in pulse distortion caused by increasing r, since the RMAs with low values of r show greater variability in the H-plane, although the SATF is stable.

Chapter 9

Printed Circuit Monopoles

Daniel Valderas and Juan I. Sancho
CEIT and Tecnun, University of Navarra

When a PMA is flattened so that its ground plane is coplanar with the radiating element instead of perpendicular to it, the result is a completely full 2D monopole antenna. This chapter demonstrates that the current distribution under specific feed conditions complies with the requirements laid down for applying the transmission line model, and then goes on to demonstrate that the staircase profile is still useful for defining the operating range. As in Chap. 5, three prototypes are taken as cases for study. These were designed to have increasing operating ranges while maintaining the same dimensions. The chapter ends by comparing the three in terms of radiation stability.

9.1. Introduction

If a UWB antenna is to be fitted in a mobile terminal device, e.g., laptop computers or mobile phones, it must have a very low profile (Wong *et al.*, 2004a; Wong *et al.*, 2004c). As increasing degrees of integration are required, printed circuit board (PCB) antennas must be used. These are eminently practical, as they can be fitted in very slim terminal devices. In the case of monopole antennas, the resulting one is known as a printed circuit monopole (PCM). Techniques for increasing impedance bandwidth that are similar to those for 3D PMAs can be also applied in this case. Examples are staircase-profile implementation (Lee *et al.*, 2005; Rambabu *et al.*, 2006) and ground plane-monopole transition taper (Eldek, 2006; Liu and Kao, 2005; Jung *et al.*, 2008). When an antenna forms part of a PCB, there are few options for making the antenna more omnidirectional or stable over frequency. In this case, the correct trade-off must be found between the "low-profile" characteristic and the loss of radiation stability due to its planar structure. With this aim in mind, how an increase in bandwidth affects the radiation stability of a UWB PCM should also be investigated for practical applications.

9.2. Current Distribution in a PCM

Figure 9.1 shows a PCM and its design parameters (Valderas *et al.*, 2008). A hole could be made between the staircase profile and the ground plane, so as to avoid the dielectric affecting the prediction of the added resonances.

As shown in the figure, a probe feed has been placed as close as possible (3 mm) to the monopole sheet. This is to avoid the negative effect on matching caused by an antenna feed via a (typically long) microstrip or printed coplanar waveguide (CPW). This antenna thus resembles those in previous chapters, except for the fact that it has been flattened. A different profile, e.g., bevel, could have been implemented instead of the staircase profile.

The current distribution is shown in Fig. 9.2 and corroborates transmission line behaviour around the antenna feed. It can be observed that the horizontal current distribution predominates over the vertical along the lower edge of the monopole sheet and upper edge of the ground plane. The circled area of the radiator in the figure can therefore be considered to act as a transmission line. The previously studied criteria are applicable.

9.3. TLM Applied to a PCM

As the current distribution is that of a transmission line, the resonances introduced by a staircase profile will follow the same principle described in Chap. 5. According to that chapter, an estimated maximum upper VSWR < 2 impedance bandwidth limit of f_u for a staircase PMA is given by Eq. (5.7). This equation, restated here and now referred to as Eq. (9.1), can be applied to this case of a staircase-profile PCM with a probe feed.

$$f_u \approx n\frac{0.3}{2W}, \quad n = 1, 2, \ldots, \tag{9.1}$$

with

$$lx = \frac{W}{n + 1}, \tag{9.2}$$

where f_u is given in GHz; n is a given number of pairs of notches and W the semi-width (in metres). The parameter lx is the same for all the notches.

Equation (5.7) was defined in a dielectric-free environment. It is known that a dielectric-loaded transmission line exhibits a different value of Zo, and the new

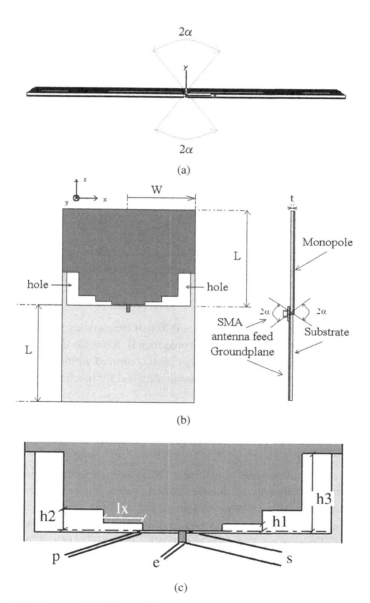

Fig. 9.1.   (a) Staircase-profile PCM top, (b) front and side view and (c) detailed view with parameters. (©2008 IEEE.)

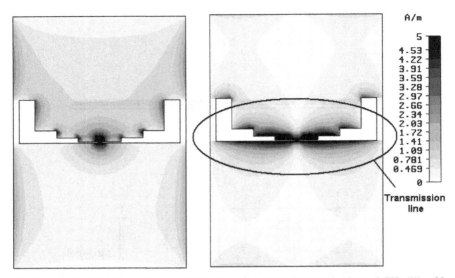

Fig. 9.2. Typical PCM (a) vertical and (b) horizontal current density distribution at 3 GHz ($W = 30$ mm, $L = 45$ mm). (©2008 IEEE.)

resonances would therefore be placed at different frequencies. This would make the design more unpredictable (but still broadband). Since the dielectric substrate is only a platform for the antenna, it may thus be removed within the transmission line zone. Two dielectric strips are, however, retained for mechanical support.

9.4. Case Study: Tailored Bandwidth for UWB PCM

9.4.1. *Design*

In Chap. 5, several increasingly broadband PMAs were designed having the same overall L and W. This implied that these PMAs had a common lower frequency limit and increasing numbers of steps in the staircase profile. Similarly, three PCMs can now be designed that have the same L and W parameters and increasing upper-frequency f_u limits, e.g., 4.85 GHz (DS-UWB), 10.6 GHz (FCC-UWB) and 15 GHz (Fig. 9.1(b)). In this way, the lower impedance bandwidth limit fl remains basically constant ($L = 45$ mm and $W = 30$ mm). This limit is estimated to be below 1.4 GHz. An estimate is used because formula (5.5) is not applicable to a PCM, since the ground plane is no longer perpendicular to the monopole.

For these specifications, Table 9.1 provides corresponding W and n pairs for PCM_1N, PCM_2N and PCM_3N.

Table 9.1. Basic characteristics for staircase PCM with specified f_u.

Antenna	W (mm)	n	f_u(GHz)
PCM_1N	30	1	5
PCM_2N	30	2	10
PCM_3N	30	3	15

Table 9.2. Characteristics for the PCM designed, with the simulated impedance bandwidth (dimensions in mm and GHz).

Antenna	Dielectric	Holes	t	p	e	s	h_1	h_2	h_3	$fl-f_u$
PCM_1N	FR-4	No	1.5	3	4	3	10	—	—	1.1–4.79
PCM_2N	Rogers RT/duroid 5870	Yes	0.508	0.4	1.7	3	3	11	—	1.1–8.97
PCM_3N	Rogers RT/duroid 5870	Yes	0.508	0.2	1.7	3	1.4	4	15	1.17–14.7

It is now necessary to adjust the p, e and h parameters (Fig. 9.1(c)) to obtain the expected VSWR < 2 f_u (Table 9.1). Table 9.2 presents those values along with the simulated bandwidth as given by CST Microwave Studio Suite. The value of the parameter s has remained constant and as short as possible.

The laminated dielectric platform to be used depends on the frequency at which the prototype will operate. For PCM_1N, FR4 ($\varepsilon_r = 4.4$, $t = 1.5$ mm) is used. In this case, there is no need to drill the antenna. For PCM_2N and PCM_3N, which are to operate at higher frequencies, Rogers RT/duroid 5870 is preferred ($\varepsilon_r = 2.33$, $t = 0.508$ mm).

9.4.2. *Simulations and measurements*

9.4.2.1. *Impedance bandwidth*

Figure 9.3 shows the measured VSWR (Valderas *et al.*, 2008). Table 9.3 presents the measured impedance bandwidths for the three examples.

Comparing Table 9.1 and Table 9.3, the f_u limit proposed by Eq. (9.1) is shown to be a good VSWR < 2 estimate. The parameters given in Table 9.2 are not the only ones that can possibly meet the requirements. According to the Bode–Fano criterion (Fano, 1950), it is expected that there will be a trade-off in the designs between increasing the upper frequency limit and the quality of matching. Thus, for example, since PCM_3 exhibits a value of f_u that is slightly greater than the proposed limit (15 GHz), VSWR is just over 2 at certain frequencies. Likewise, PCM_2 may be designed to have a value of f_u closer to 10 GHz, but this would

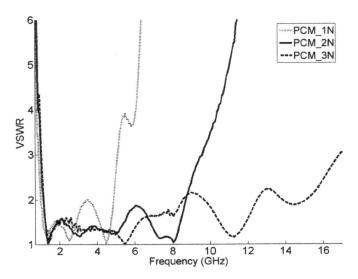

Fig. 9.3.   Measured VSWR for PCM_1N, PCM_2N and PCM_3N.

Table 9.3.   Measured VSWR < 2
impedance bandwidth for the PCM.

Antenna	Measured $f_l - f_u$ (GHz)
PCM_1N	1–4.87
PCM_2N	1.1–8.7
PCM_3N	1.15–15.15[a]

imply increased computing and prototyping costs. This approach is therefore a useful guideline and gives an indication of what can be obtained by subsequent optimisation algorithms.

9.4.2.2. *Antenna gain*

For the axes in Fig. 9.1, the radiation patterns are mainly symmetrical with respect to the XZ and YZ planes. Gain and radiation patterns are kept mainly constant as increasing numbers of steps or notches are implemented in the antenna. This is shown in Fig. 9.4, where maximum gain (IEEE) is computed for the three antennas within their respective impedance bandwidth. At lower frequencies, the antennas

---

[a]Over 8.6–9.6 GHz and 12.5–13.7 GHz, 2 < VSWR < 2.25.

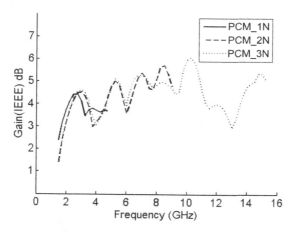

Fig. 9.4.   Absolute PCM gain within the impedance bandwidth. (©2008 IEEE.)

exhibit low gain, as they are still electrically small. Maximum gain is observed in the y direction for frequencies up to 4 GHz ($2W \approx \lambda/2$) (Wong *et al.*, 2005b). As explained in Chap. 7 for the concept of CRO, above this frequency, the maximum radiation direction changes. Gain for such frequencies is chiefly between 3.5 and 5.5 dB.

9.4.2.3. *Angular range computation*

These increasingly broadband PCMs having been designed, it is assumed that the values of SATF and SAGD will be similar over the same VSWR < 2 impedance bandwidth and, as in Fig 5.30, a filtering effect in the out of band region will be observed. To translate this situation into the coverage for a UWB handset, the concept introduced in Chap. 2, *Angular Range* $A(\theta_o, \Phi o)$, is brought into play. It defines the maximum angle of coverage for which the radiated signal is stable (PSF > 0.95), measured from a reference direction (θ_0, Φ_0), on a plane and over a specific bandwidth. This angular range is now studied, in order to compare the different PCM antennas.

In the study, it is assumed that the z and x axes are not relevant directions. E_θ fields over the E-plane and H-plane are calculated at $5°$ intervals on both sides of the antenna, taking the broadside $\pm y$ axis ($\theta = 90°$, $\Phi = 90°, 270°$) as a reference. CST Microwave Studio Suite was used to calculate these radiated fields for transmission at 1,000 mm, in far-field conditions as a Gaussian pulse response. The PSF parameter, defined in Chap. 2 (Dissanayake and Esselle, 2006), can be computed from 0–15 GHz frequency domain transfer functions for any 2α angle

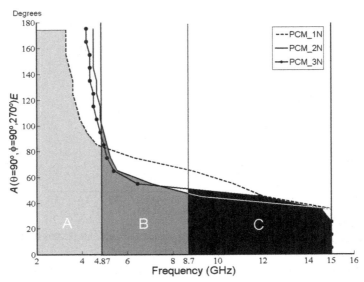

Fig. 9.5. $A(\theta = 90°, \Phi = 90°, 270°)_E$ versus frequency and operation range for the PCMs. (©2008 IEEE.)

of coverage (Fig. 9.1). For each $A(\theta = 90°, \Phi = 90°, 270°)$ with PSF > 0.95, an upper operating frequency is found. As the $A(\theta = 90°, \Phi = 90°, 270°)$ is decreased, this frequency increases to beyond the upper impedance bandwidth limit frequency, given by Eq. (9.1).

Figure 9.5 shows $A(\theta = 90°, \Phi = 90°, 270°)_E$ (E-plane) versus frequency for the three prototypes. The trend of the plot is seen to decrease with frequency in all cases: at the higher bandwidths under consideration, it is observed that the transmission signals are correlated over smaller angles. The operating range area (e.g., A-zone for PCM_1N) is bounded by the plot and the corresponding upper impedance bandwidth limit frequency (4.87 GHz in the example). Areas B and C are additional zones that have been added for PCM_2N and PCM_3N, in that order. Likewise, Fig. 9.6 shows $A(\theta = 90°, \Phi = 90°, 270°)_H$ (H-plane) versus frequency.

Figures 9.5 and 9.6 indicate that the PCM_1N exhibits coverage up to 83° in the E-plane and 115° in the H-plane on both sides of the antenna, with an upper frequency limit of 4.85 GHz DS-UWB. As the antenna has a VSWR < 2 lower frequency limit of 1 GHz, it may also be used for DCS 1800, PCS 1900, WCDMA, 3G, 802.11a/b/g, Bluetooth®, ZigBee®, WiMAX™, etc. PCM_2N and PCM_3N also provide that coverage and, for FCC-compliant 10.6 GHz UWB operation,

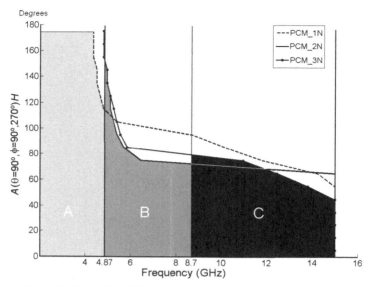

Fig. 9.6. $A(\theta = 90°, \Phi = 90°, 270°)_H$ versus frequency and range of operation for the PCMs. (©2008 IEEE.)

they feature stable coverage up to approximately 45° in the E-plane and 73° in the H-plane. The PCM_3N is suitable for further direction-specific applications at frequencies below 15 GHz.

Consequently, as higher frequencies are required for UWB applications, PCMs can be synthesised by increasing the number of steps in the antenna's profile while keeping the same length and width in order to determine a fixed lower limit frequency. Thus, each PCM performs similarly to the previous ones, but the operating bandwidth is extended to higher frequencies where the radiation is stable over a narrower angular range. This property makes them suitable for application-specific UWB communications links.

Chapter 10

Applications of UWB Antennas

Xiaodong Chen
Queen Mary, University of London

As it was introduced in Chap. 1, UWB antennas have been widely applied to wireless communications, radars, imaging, localisation and wireless sensor networks. In this chapter, we only highlight some UWB antennas and their latest developments in applications to UWB communications, electromagnetic (EM) testing, medical imaging and radars due to space limitation. In practical applications, apart from the requirements of a good EM performance, a UWB antenna has to account for many other demands, such as small size, easiness to be integrated with the system and mechanical robustness, etc. Hence, the final choice of UWB antennas in various applications is always a trade-off between different requirements.

10.1. UWB Communications

As mentioned in Chap. 1, UWB communications have split into two modulation schemes: impulse radio or Direct Sequence — UWB (DS-UWB) and Multiband OFDM (MB-OFDM).

In the impulse radio scheme, the ultra wide bandwidth is required for the communication system. In North America, 3.1–10.6 GHz, i.e., 7500 MHz is allowed by FCC, while in Europe and other countries, two sub-bands, i.e., 3.1–4.8 GHz and 6.2–10.6 GHz can be exploited. Therefore, the antennas used for this scheme need to have good impedance matching and consistent radiation property, as well as a good time-domain characteristic (small group delay) over the corresponding bandwidth.

In the Multiband-OFDM scheme, the 3.1–10.6 GHz UWB band is further divided into 14 sub-bands, 528 MHz each. The multiband-OFDM communication system only utilises a 528 MHz channel at a time. So, the demand on the antenna design can be relaxed. For example, a good time-domain characteristic is no longer required. Sometimes, the antenna is only required to operate over one sub-band, say, 3.1–4.8 GHz.

10.1.1. *Antennas required in impulse radio system*

The compact and small UWB antenna required for the impulse radio system was the main drive for the UWB antenna research a few years ago. Vast numbers of UWB antennas have been developed for this application worldwide. Broadly speaking, these antennas can be divided into three categories, namely, dipoles, monopoles and others. Here, we examine the evolution of UWB antennas being employed in the commercial impulse radio systems.

The first generation of impulse radio systems all utilise the dipole type of UWB antennas, such as Time Domain's PulseOn 200 antenna (Time Domain, 2003) and Artimi's antenna (Guo, 2006).

The PulseOn 200 antenna is printed elliptical dipole with an approximate dimension of 40×70 mm, as shown in Fig. 10.1. The upper arm is printed on both sides of the PCB, being connected via holes at the edge of the ellipse. The feed is a tapered twinline, connecting the lower arm through a slot and the upper arm on the opposite side. Actually, this antenna was only designed to operate between 3.0 and 5.5 GHz.

Fig. 10.1.   Time Domain's PulseOn 200 impulse radio unit and the antenna. (Time Domain, 2003.)

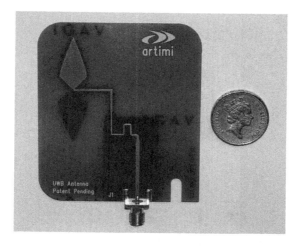

Fig. 10.2.   Arimi's UWB dipole antenna. (Guo, 2006.)

Another UWB dipole example is the Artimi's antenna, as shown in Fig. 10.2 (Guo, 2006). Each arm of the dipole is a flair, printed on the opposite sides of a substrate. The feed is a twinline connected to a balun transformer to the SMA connector. The antenna has a good impedance matching (-10 dB return loss) over FCC defined UWB band: 3.1–10.6 GHz and exhibits good time domain characteristics (Guo, 2006).

The UWB dipole antennas are normally operated with the impulse radio system as a standalone unit due to their big size (see Fig. 10.1). It is desirable to have small UWB antennas which can be mounted/integrated on a PCB. Therefore, considerable efforts have been spent on the miniaturisation of the UWB antennas. Printed UWB monopoles become a favorable choice as they do not require a balanced feed and can be integrated with the radio unit. Here are two UWB monopole examples.

SkyCross SMT-3TO10M antenna is stamped metal monopole on PCB with a self-contained matching structure, as shown in Fig. 10.3 (SkyCross, 2006). It has an impedance bandwidth (VSWR: 2.0:1) between 3.6–9.1 GHz and a dimension of 26.2×18.5 mm. It is commercially available and has been used in many impulse radio systems.

A more recent monople type of UWB antenna is a chip antenna (3100AT51A7200) based on the LTCC technology developed at Johanson Technology, USA. The antenna element has a dimension of 6 mm $\times$ 10 mm and can be easily integrated into a PCB, as shown in Fig. 10.4 (Johanson Technology, 2006). It has an impedance bandwidth: 3.1–10.3 GHz, measured at 10 dB return

Fig. 10.3.   SkyCross SMT-3TO10M antenna. (SkyCross, 2006.)

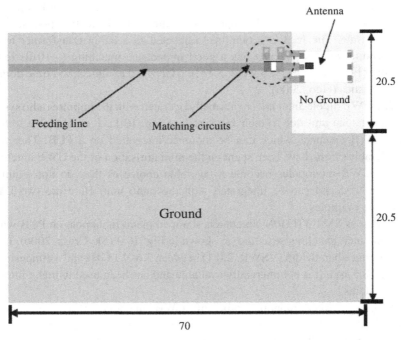

Fig. 10.4.   Johanson Technology's UWB chip antenna and layout on a PCB. (Johanson Technology, 2006.)

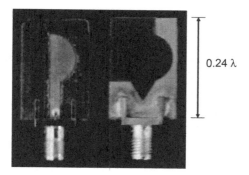

0.24 λ

Fig. 10.5. Front and back views of QMUL quasi-self-complementary UWB antenna. (Guo, 2009.)

loss and an average gain of −3.5 dBi. The antenna is commercially available for the deployment in impulse radio systems.

Apart from these UWB dipoles and monopoles, there exist other types of UWB antennas, such as tapered slots, spirals and self-complementary structures. Generally speaking, these types of antennas are quite big in size and not suitable for the integration with the portable impulse radio system. However, the antenna group at Queen Mary, University of London has developed a novel compact quasi-self-complementary antenna, as shown in Fig. 10.5 (Guo, 2009). A half circular disc with its complementary magnetic counterpart (half slot) are printed on the different side of the dielectric substrate. The feed is a microstrip line with a triangular notch on the ground plane to improve the impedance matching. This antenna can offer an ultra wide return loss bandwidth (VSWR < 2.0:2.89–10.7 GHz) with reasonable radiation properties. It features a small physical size of 16 mm × 25 mm, corresponding to an electrically size of 0.24 λ.

10.1.2. *Antennas required in MB-OFDM system*

Lately, the MB-OFDM scheme has been widely adopted in the UWB communication systems, especially the Wireless USB dongles. Several versions of small UWB antennas have been developed for this application. One example is a band-notched UWB dongle antenna, as shown in Fig. 10.6 (Jung, 2008). The antenna element is a U-shaped ring monopole originally covering 3.1–10.6 GHz bandwidth. A pair of stubs is inserted in the middle of the ring to provide a rejection band around 5.6 GHz (WLAN bands). The antenna occupies an area of 20 mm × 14.5 mm, plus a 55 mm long ground plane. The antenna provides a good impedance matching between 3.1–5.15 GHz and 5.85–10.6 GHz.

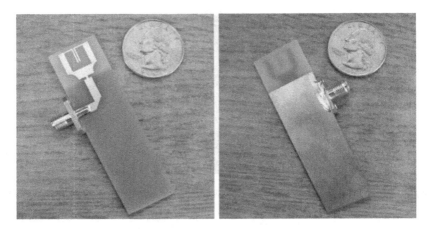

Fig. 10.6.   A band-notch UWB antenna for wireless USB dongle. (Jung, 2008.)

Fig. 10.7.   Tiny wireless USB dongle antenna. (See, 2008.)

Another example is a small wireless USB dongle antenna developed at the Institute for Infocomm Research, Singapore, as shown in Fig. 10.7 (See, 2008). The radiator is a triangle monopole with an extended wire for impedance matching and a slot cut at the top for loading. On the back, two perpendicular strips were extended from the ground plane to further tune the impedance matching. The antenna is only designed to cover the band from 3.1 GHz to 5.0 GHz, with a size of 11 mm × 20.5 mm.

10.2. EM Measurement

Historically, one important application of UWB antennas is the EM measurements for various purposes, such as antenna characterisation, EMI testing and EM spectrum monitoring. Due to the wide band nature of signal to be detected and measured, the detecting antenna needs to be broadband or ultra wideband. For example, a double ridged horn.

With a relative bandwidth 8:1 is commonly utilised in antenna radiation pattern or gain measurement. A broadband dicone is usually employed for the monitoring of the EM spectrum. A broadband log-periodic antenna is normally used to capture the EM emission from the electronic devices for EMC testing. Most of these antennas are well documented in various textbooks or literatures. Here, we only introduce some new developments in EM measurement antennas.

10.2.1. *EM spectrum monitoring*

Recently, a compact UWB antenna subsystem has been developed at Queen Mary, University of London, for the Autonomous Interference Monitoring System (AIMS), which was commissioned by Ofcom (Office of Communications, UK) to monitor EM spectrum usage in the urban, suburban and rural locations. The UWB antenna for the AIMS system is required to have omni-directional radiation patterns and a bandwidth (VSWR < 2–3) from 100 MHz to 10.6 GHz. Simultaneously, the antenna is also required to be dual polarised across most of the frequency range.

The UWB antenna subsystem developed consists of two UWB antennas covering the low and high bands, respectively, as shown in Fig. 10.8 (Chiau, 2006). The antennas are printed on a FR4 substrate having a dimension of 900 mm × 910 mm × 1.6 mm. The low band UWB antenna operating over the frequency range from 100 MHz to 3 GHz is a modified version of co-planar waveguide (CPW) fed circular ring monopole. The top and side edges of the circular ring monopole have been removed to distort the current distribution on the antenna so that it can receive dual polarised RF signals, in particular, horizontal polarised TV signals around 400 MHz. The high band UWB antenna from 3 GHz to 10.6 GHz in the UWB antenna sub-system is a CPW elliptical slot antenna. A RF switch is used to switch between these antennas when scanning the spectrum.

10.2.2. *EMC testing*

Another recent development of antenna detecting at Queen Mary, University of London, is related to the measurement of EM interference emitted from camera, display and other units inside a mobile phone handset. However, these interference

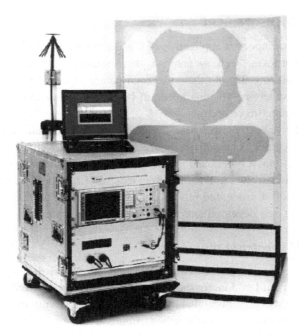

Fig. 10.8. Autonomous interference monitoring system and the detecting antenna. (Chiau, 2006.)

signals are very low in strength and can only be detected by a specially designed EMC detecting system. The developed EMC testing system and its detecting antenna are shown in Fig. 10.9 (Chen, 2008). The broadband detecting antenna is an elliptical slot fed through semi-circular ring and its bandwidth ranges from 750 MHz to 6 GHz (VSWR < 2.0). The EMC detecting system is very sensitive and can detect the interference signal as low as −135 dBm using a broadband LNA.

10.3. Medical Imaging — Breast Cancer Detection

It is recently noticed that small antennas are required for some medical applications (Chen, 2008). One of the most exciting medical applications which require the UWB antennas is microwave imaging for breast cancer. This imaging modality transmits lower power short microwave pulses (tens picoseconds to nanoseconds) into the body and detects the differentiated scattering due to the dielectric contrast between malignant and healthy tissues. An UWB compact antenna array is employed to conduct a 3D scan through a so-called digital beamforming in space-time domain (Li, 2005). In order to achieve high detection resolutions and good clutter rejection, the antennas with an impedance bandwidth 4–10 GHz are usually

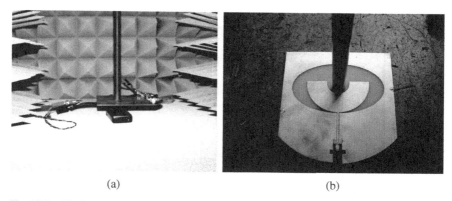

(a)   (b)

Fig. 10.9.   EMC testing unit for mobile terminal (a) and the antenna (b), developed at Queen Mary, University of London.

employed. Both omni-directional and directional types of small UWB antennas have been developed for this application worldwide. Lately, the research has been focused on the directional type of UWB antennas. Here we review a number of these designs.

10.3.1. *Horn and tapered slot antennas*

The first small UWB antenna for breast cancer imaging application was studied by the University of Wisconsin-Madison group (Li, 2005). The antenna is a modified pyramidal horn, with a metallic ridge connecting one side wall of the horn and the outer conductor of the coaxial feed, and a curved metallic launching plane connecting the central conductor of the coaxial feed. It is terminated on the other side wall of the horn with two 100 Ω chip resistors, as shown in Fig. 10.10. The pyramidal horn features a small size, with a depth of 13 mm and a 25 mm $\times$ 20 mm aperture. A good impedance matching has been achieved over the operating frequency band: 2–11 GHz. The fidelity of the antenna is found to be very good, greater than 0.92 in most directions. However, the drawback of this antenna is the low efficiency due to the resistive loading on the launching plane.

The tapered slot antenna (TSA) was also proposed to be used for this application. One of the early designs was studied by the University of Queensland group (Khor, 2007), as shown in Fig. 10.11. The antenna has a -10 dB return loss over bandwidth 2.75–11 GHz with a size 50 mm $\times$ 50 mm.

A modified antipodal antenna was also designed at Queen Mary, University of London, for this application. The antenna has both round arms and smooth transactions to the feed and ground plane, respectively, as shown in Fig. 10.12. An

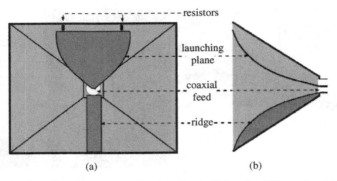

Fig. 10.10. Configuration of horn antenna developed by the University of Wisconsin-Madison group. (a) Front view and (b) Side view. (Li, 2005.)

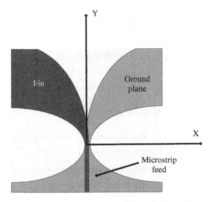

Fig. 10.11. Configuration of tapered slot antenna developed at the University of Queensland group. (Khor, 2007.)

extremely wide impedance bandwidth (2.3–20 GHz) can be obtained with a size 50 mm × 50 mm.

10.3.2. *Stacked patch antennas*

Another type of UWB antenna for breast cancer imaging is the stacked patch antenna investigated by the University of Bristol group (Klemm, 2009). Basically, it is a cavity-backed stacked-patch configuration with a microstrip line feed through a slot on the ground plane, as shown in Fig. 10.13. The antenna has a −10 dB impedance match over 4–9 GHz in the absorbing liquid. Good radiation

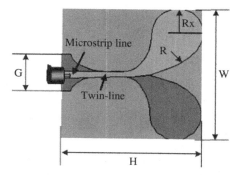

Fig. 10.12.   The antipodal antenna developed at Queen Mary, University of London. (Chen, 2008.)

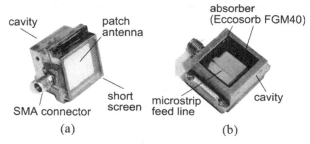

Fig. 10.13.   Cavity backed stacked patch antenna developed at the University of Bristol. (a) Front view and (b) Back view. (Klemm, 2009.)

characteristics have been obtained in simulations and measurements. Also, a suitable time-domain characteristic is shown in the measurement as well. The size of the cavity is 23 mm × 29 mm. The Bristol group has developed the first laboratory prototype of UWB imaging system for breast cancer detection, using a 4 × 4 element hemi-spherical conformal array, as shown in Fig. 10.14 (Klemm, 2009).

10.4. Radars

A radar system usually employs an antenna array for its operation. UWB arrays are also widely used in the radar systems, such as the impulse radar and the phased array radar. There are already plenty of literatures covering these subjects. Here we examine the relevant UWB antenna arrays from a different perspective and review the latest development.

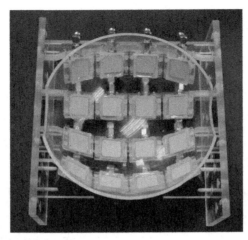

Fig. 10.14. A 4 × 4 hemi-spherical UWB antenna array developed at the University of Bristol. (Klemm, 2009.)

10.4.1. *Antenna array in impulse radars*

An impulse radar radiates the pulsed EM wave and detects the reflected pulses from the objects. It can be used as ground penetration radars (GPR) to detect buried landmines and other unexploded ordinance, through-wall radars and biomedical imaging as mentioned earlier. The antenna element required for this application needs to have a minimal dispersion of the UWB impulse, similar to that in the impulse radio.

Various UWB antenna elements have been developed to form an antenna array for impulse radars. Here we only review some typical tapered slot antenna elements.

The Vivaldi antenna is one type of tapered slot antenna and has an exponentially tapered slot for a broad bandwidth with minimal length and mismatch, as shown in Fig. 10.15(a). It is noticed at Queen Mary, University of London, that the edge of both arms can be rounded to further improve the bandwidth, as shown in Fig. 10.15(b). Normally, the feeding to a Vivaldi antenna is through a microstrip to a slot line transformer, which limits the high end of the bandwidth. Typically, a 15:1 bandwidth can be achieved in an optimised design.

A bunny ear antenna is another type of tapered slot antenna. The bunny ear antenna, as shown in Fig. 10.16, consists of a pair of tapered wings, which is much shorter than the Vivaldi antenna. The feed is like a Vivaldi antenna, using a microstrip to a slot line transformer. The bandwidth can reach 15:1 in an optimal design.

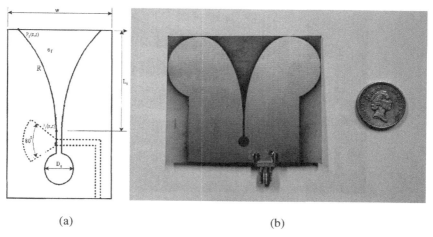

(a)   (b)

Fig. 10.15.   (a) Configuration of Vivaldi antenna, (b) A modified Vivaldi antenna developed at Queen Mary, University of London. (Chen, 2008.)

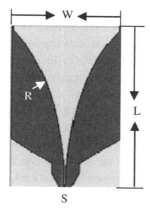

Fig. 10.16.   Typical configuration of a bunny ear antenna.

In most cases, the elements in the array are well-separated in space to minimise the coupling between antenna elements, which will degrade the array performance. An example of such an array is the stacked patch array used for breast cancer imaging, as shown in Fig. 10.16. Another example is a 4 tapered slot array for achieving a high gain over a broadband (3–10 GHz) developed at Queen Mary, University of London, is shown in Fig. 10.17.

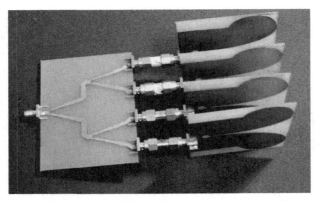

Fig. 10.17. A high gain broadband 4 elements TSA array developed at Queen Mary, University of London.

10.4.2. *Broadband phase array*

A phased array radar can steer the EM beam to scan swiftly the space electronically by adjusting the phase shifters to each antenna element. Recently, considerable research efforts have been spent on developing a broadband phase array for surveillance sensors with high resolution, long-range imaging capabilities and high sensitivity. Thus, it has a great potential for ground penetration detection, through-wall imaging, surveillance/security for high-security premise application and medical imaging. However, designing a compact broadband array is very challenging due to the conflicting requirements posed by the effect of inter-element spacing. To avoid grating lobes at the high frequency end of the band, the spacing between each element needs to be minimal. But the close spacing will result in a strong mutual coupling between the elements, which degrades impedance bandwidth and performance.

The most popular antenna arrays being studied by many researchers are TSA arrays, as discussed in the previous section. This kind of array is relatively easy to be fabricated using printed circuit technique and is conveniently fed.

One example is a test array of 25×25 bunny ear elements, as shown in Fig. 10.18 (Lee, 2004). The bunny ear antenna is shown in Fig. 10.16. It works well over 3–14 GHz with a low profile of 12 mm.

Another example is a doubly mirrored balanced antipodal Vivaldi (DmAVA) antenna array, as shown in Fig. 10.19 (Schubert, 2006). By employing mirror symmetries in the array of balanced antipodal Vivaldi antennas, most of the impedance anomalies introduced by the gaps between elements can be removed.

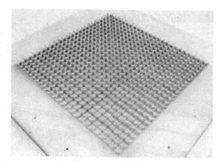

Fig. 10.18. A 25 × 25 element X-band array using bunny-ear elements for 3–14 GHz demonstration. (Lee, 2004.)

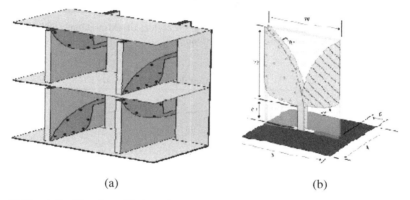

(a)   (b)

Fig. 10.19. A Doubly mirrored balanced Antipodal Vivaldi antenna (DmBAVA) as a modular array structure (Schubert, 2006): (a) Sketch of the array unit cell (b) Structure of the BAVA unit cell.

The DmAVA array was demonstrated through simulations to operate well over more than one octave with a scan angle up to 45 degrees.

In order to further increase the operation bandwidth and the scanning angle, a new design concept was introduced at Queen Mary, University of London. Instead of trying to suppress the mutual coupling, a compact TSA array was designed based on tightly-spaced elements which resulted in a strong mutual coupling between elements. This makes it possible to radiate effectively at a lower frequency band due to the radiation from adjacent elements, which in turn retains a good performance at a higher frequency band (Wang, 2008).

A compact prototype of 32 element TSA linear array (32 cm long) was designed at Queen Mary, University of London, as shown in Fig. 10.20. A broad bandwidth of 4–18 GHz with a scanning angle up to 60 degrees is achieved by controlling

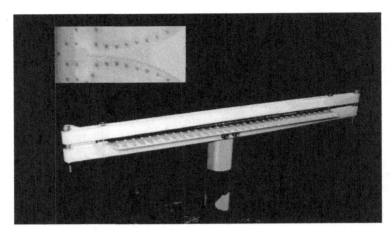

Fig. 10.20.   Photo of broadband 32 TSA element array (32 cm long) based on strongly mutual coupling (Insert is the photo of the TSA element: 9 mm × 31 mm).

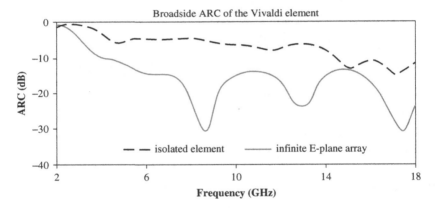

Fig. 10.21.   Active reflection coefficient curves of an isolate element and an infinite array.

the mutual coupling between the elements. The S11 curves of an isolated element and an infinite planar array are shown in Fig. 10.21. It is interesting to note that the bandwidth of the element is very narrow at 18 GHz, but the array bandwidth is substantially broadened, benefitting from the element mutual coupling which acts as a reactive loading.

Bibliography

Agrawall, N. P., Kumar, G. and Ray, K. P. (1998). Wide-band planar monopole antennas. *IEEE Transactions on Antennas and Propagation.* 46:2, 294–295.

Aiello, R. and Batra, A. (2006). *Ultra Wideband Systems: Technologies and Applications.* Elsevier, Oxford.

Alsawaha, H. and Safaai-Jazi, A. (2009). New design for Ultra-Wideband hemispherical helical antennas. *Antennas and Propagation Society International Symposium.*

Amert, A. K. and Whites, K. W. (2009). Miniaturization of the biconical antenna for ultra-wideband applications. *IEEE Transactions on Antennas and Propagation.* 57:12, 3728–3735.

Ammann, M. J. (2000). Impedance bandwidth of the square planar monopole. *Microwave and Optical Technology Letters.* 24:3, 185–187.

Ammann, M. J. (2001). Control of the impedance bandwidth of wideband planar monopole antennas using a beveling technique. *Microwave and Optical Technology Letters.* 30:4, 229–232.

Ammann, M. J. and Chen, Z. N. (2004). An asymmetrical feed arrangement for improved impedance bandwidth of planar monopole antennas. *Microwave and Optical Technology Letters.* 40:2, 156–158.

Ammann, M. J. and Cordoba, R. S. (2004). On pattern stability of the crossed planar monopole. *Microwave and Optical Technology Letters.* 40:4, 294–296.

Anob, P. V., Ray, K. P. and Kumar, G. (2001). Wideband orthogonal square monopole antennas with semi-circular base. *Proceedings of the International Symposium of Antennas and Propagation.* 3, 294–296.

Antonino-Daviu, E., Cabedo-Fabres, M., Ferrando-Bataller, M. and Valero-Nogueira, A. (2003). Wideband double-fed planar monopole antennas. *Electronics Letters.* 39:23, 1635–1636.

Arai, H. (2001). *Measurement of Mobile Antenna Systems.* Artech House, Boston.

Arslan, H., Chen, Z. N. and Benedetto, M. D. (2006). *Ultra Wideband Wireless Communication.* John Wiley & Sons, New Jersey.

Balanis, C. A. (1997) *Antena Theory, Analysis and Design.* Wiley, New York.

Baum, C. E. and Farr, E. G. (1999). *Impulse Radiating Antennas.* Plenum Press, New York.

Brown, G. H. and Woodward, O. M. (1945). Experimentally determined impedance characteristics of cylindrical antennas, *Proceedings of the I.R.E.* pp. 257–262.

Cabedo-Fabres, M., Antonino-Daviu, E., Valero-Nogueira A. and Ferrando-Bataller, M. (2003). Analysis of wide band planar monopole antennas using characteristic modes. *IEEE Antennas and Propagation Society International Symposium.* 3, 733–736.

Cardama, A., Jofre, L., Rius, J. M., Romeu, J. and Blanch, S. (1998). *Antenas.* Ediciones UPC, Barcelona.

Chen, X., Gao, Y., Guo, L. and Dupuy, J. (2008). EMC Testing Apparatus, European Patent Application No. 08100712.2.

Chen, X., Wang, Z., Wang, S. and Parini, C. G. (2008). Small UWB antennas for medical applications. *LAPC 2008*, 2008, Loughborough, UK, pp. 28–31.

Chen, Z. N. (2000a). Impedance characteristics of planar bow-tie-like monopole antennas. *Electronics Letters*. 36:13, 1100–1101.

Chen, Z. N. (2000b). Broadband planar monopole antenna. *IEE Proc.–Microwaves, Antennas and Propagation*. 147:6, 526–528.

Chen, Z. N. (2005). Novel bi-arm rolled monopole for UWB applications, *IEEE Transactions on Antennas and Propagation*. 53:2, 672–677.

Chen, Z. N. and Chia, Y. W. M. (2000a). Impedance characteristics of trapezoidal planar monopole antennas. *Microwave and Optical Technology Letters*. 27:2, 120–122.

Chen, Z. N. and Chia, Y. W. M. (2000b). Broadband monopole antenna with parasitic planar element. *Microwave and Optical Technology Letters*. 27:3, 209–210.

Chen, Z. N. and Chia, Y. W. M. (2001). Impedance characteristics of EMC triangular planar monopoles. *Electronics Letters*. 37:21, 1271–1272.

Chen, Z. N, Ammann, M. J., Chia, M. Y. W. and See, T. S. P. (2002). Annular planar monopole antennas. *IEE Proc.–Microwave, Antennas and Propagation*. 149:4, 200–203.

Chen, Z. N., Ammann, M. J. and Chia, M. Y. W. (2003). Broadband square annular planar monopoles. *Microwave and Optical Technology Letters*. 36:6, 449–454.

Chen, Z. N., Wu, X. H., Li, H. F., Yang, N. and Chia, M. Y. W. (2004). Considerations for source pulses and antennas in UWB radio systems. *IEEE Transactions on Antennas and Propagation*. 52:7, 1739–1748.

Chiau, C. C., Dupuy, J., Chen, X. and Parini, C. G. (2006). Design of UWB Antenna for Autonomous Interference Monitoring System (AIMS). *Antennas and Propagation Society International Symposium 2006, IEEE*. pp. 4633–4636.

Collin, R. E. (1956). The optimum tapered transmission line matching section. *Proceedings of IRE*. 44, 539–548.

Daviu, E. A., Cabedo Fabrés, M., Ferrando Bataller, M. V. and Nogueira, A. (2003). Wideband double-fed planar monopole antennas. *Electronic Letters*. 39:23, 1635–1636.

Dissanayake, T. and Esselle, K. P. (2006). Correlation-based pattern stability analysis and a figure of merit for UWB antennas. *IEEE Transactions and Antennas Propagation*. 54:11, 3184–3191.

Dou, W. and Chia, Y. W. M. (2000). Small broadband stacked planar monopole. *Microwave and Optical Technology Letters*. 27:4, 288–289.

Dubost, G. S. and Zisler (1976). *Antennas a large bande*. Masson, New York.

DuHamel, R. and Isbell, D. (1957). Broadband logarithmically periodic antenna structures. *IRE International Convention Record*. 5:1, 119–128.

Dyson, J. D. (1959). The unidirectional equiangular spiral antenna. *IRE Transactions on Antennas and Propagation*. 7:4, 329–334.

Eldek, A. (2006). Numerical analysis of a small ultrawideband microstrip-fed tap monopole antenna. *Progress In Electromagneticism Research*. 65, 59–69.

Evans, J. A. and Ammann, M. J. (2003). Planar monopole design considerations based on TLM estimation of current density. *Microwave and Optical Technology Letters*. 36:1, 40–42.

Fano, R. M. (1950). Theoretical limitations on the broadband matching of arbitrary impedances. *Journal of Franklin Institute.* 249, 139–154.

Federal Communications Commission (FCC) (2002). Revision of Part 15 of the commission's rules regarding ultra-wideband transmission systems, First Report and Order, pp. 98–153. ET Docket.

http://www.fcc.gov/Bureaus/Engineering_Technology/News_Releases/2002/nret0203.ppt #2

Garg, R., Bhartia, P., Bahl, I. and Ittipiboon, A. (2000). *Microstrip Antenna Design Handbook.* Artech House, Boston.

Ghavami, M., Michael, L. B. and Kohno, R. (2008). *Ultra Wideband Signals and Systems in Communication Engineering,* 2nd ed. Wiley, New York.

Greenberg, M. C., Virga, K. L. and Hammond, C. L. (2003). Performance characteristics of the Dual Exponentially Tapered Slot Antenna (DETSA) for wireless communications applications. *IEEE Transactions on Vehicular Technology.* 52:2, 305–312.

Griffin, J. D. and Durgin, G. D. (2009). Complete link budgets for backscatter radio and RFID systems. *IEEE Antennas and Propagation Magazine.* 51:2. 11–25.

Guo, L., Liang, J., Chen, X. and Parini, C. G. (2006). Time domain behaviors of Artimi's UWB antenna, 2006. *IEEE International Workshop on Antenna Technology, Small Antennas and Novel Metamaterials.* pp. 297–300.

Guo, L., Wang, S., Chen, X. and Parini, C. G. (2009). A small printed quasi-self-complementary antenna for ultrawideband systems. *IEEE Antennas and Wireless Propagation Letters.* 8, 554–557.

G. Lu, von der Mark, S., Korisch, I., Greenstein, L. J. and Spasojevic, P. (2004). Diamond and rounded diamond antennas for ultrawide-band communications. *IEEE Antennas and Wireless Propagation Letters.* 3:13, 249–252.

Hallen, E. (1938). Theoretical investigations into transmitting and receiving qualities of antennae. *Nova Acta Regiae Soc. Sci. Upsaliensis, Ser.IV.* 11:1.

Han, S. H. and Lee, J. H. (2005). An overview of peak-to-average power ratio reduction techniques for multi-carrier transmission. *IEEE Wireless Communication.* 12, 56–65.

Honda, S., Ito, M., Seki, H. and Jingo, Y. (1992). A disc manspole with 1:8 impedance bandwidth and omnidirectional radiation pattern. *Proceedings of the International symposium of Antennas Propagation.* Sapporo, Japan. pp. 1145–1148.

Hood, A. Z., Karacolak, T. and Topsakal, E. (2008). A small antipodal Vivaldi antenna for Ultrawide-Band applications. *IEEE Antennas and Wireless Propagation Letters.* 7, 656–660.

Jordan, E. C., Deschamps, G. A. and Dyson, J. D. (1964). Developments in broadbands antennas. *IEEE Spectrum.* 1, 58–71.

Jung, J., Seol, K., Choi, W. and Choi, J. (2005). Wideband monopole antenna for various mobile communication applications. *Electronics Letters.* 41:24, 1313–1314.

Jung, J., Choi, W. and Choi, J. (2005). A small wideband microstrip-fed monopole antenna. *IEEE Transactions of Antennas and Propagation.* 15:10, 703–705.

Kerkhoff, A. and Ling, H. (2002). The design and analysis of miniaturized planar monopoles. *IEEE Antennas and Propagation Society International Symposium.* 4, 30–33.

Kerkhoff, A. and Ling, H. (2003). Design of a planar monopole antenna for use with ultra-wideband (UWB) having a band-notched characteristic. *IEEE Antennas and Propagation Society International Symposium.* 1, 830–833.

Kerkhoff, A., Rogers, R. and Ling, H. (2001). The use of the genetic algorithm approach in the design of ultra-wideband antennas. *IEEE Radio and Wireless Conference.* pp. 93–96.

Kerkhoff, A. J., Rogers, R. L. and Ling, H. (2004). Design and analysis of planar monopole antennas using a genetic algorithm approach. *IEEE Transactions on Antennas and Propagation.* 52:10, 2709–2718.

Khor, W. C., Bialkowski, M. E., Abbosh, A., Seman, N. and Crozier, S. (2007). An ultra wideband microwave imaging system for breast cancer detection. *IEICE Transactions of Communications.* E90-B(9), 2376–2381.

Kim, K. H., Kim, J. and Park, S. O., (2005). An ultrawide-band double discone antenna with the tapered cylindrical wires. *IEEE Transactions on Antennas and Propagation.* 53:10, 3403–3406.

Kiminami, K., Hirata, A. and Shiozawa, T. (2004). Double-sided printed bow-tie antenna for UWB communications. *IEEE Antennas and Wireless Propagation Letters.* 3:9, 152–153.

Klemm, M. and Tröster, G. (2005). Characterization of small planar antennas for UWB mobile terminals. *Wireless Communications and Mobile Computing.* 5, 525–536.

Klemm, M., Craddock, I. J., Leendertz, J. A., Preece, A. and Benjamin, R. (2009). Radar-based breast cancer detection using a hemispherical antenna array — experimental results. *IEEE Transactions on Antenna and Propagation.* 57:6, 1692–1704.

Klopfenstein, R. W. (1956). A transmission line taper of improved design. *Proceedings of IRE.* 44, 15–31.

Kraus, J. D. (1950). *Antennas.* McGraw-Hill, New York.

Kumar, G. and Ray, K. P. (2003). *Broadband Microstrip Antennas.* Artech House. Boston.

Kumar, N. and Buehrer, R. M. (2008). The ultra wideband WiMedia standard. *IEEE Signal Process.* 25, 115–119.

Johanson Technology Inc (2006). Ultra wide band (UWB) chip antenna 3.1–10.3 GHz, detailed specification.

Jung, J., Lee, H. and Lim, Y. (2008). Band notched ultra wideband internal antenna for usb dongle application. *Microwave and Optical Technology Letters.* 50:7, 1789–1793.

Lee, E., Hall, P. S. and Gardner, P. (1999). Compact wideband planar monopole antenna. *Electronic Letters.* 35:25, 2157–2158.

Lee, J. J., Livingston, S. and Koenig, R. (2004). Performance of a wideband (3–14 GHz) dual-pol array. *IEEE 2004 Antennas and Propagation Society International Symposium.* pp. 551–554.

Lee, R. T. and Smith, G. S. (2004). On the characteristic impedance of the TEM horn antenna. *IEEE Transactions on Antennas and Propagation.* 52:1, 315–318.

Lee, S. S., Choi, S. S., Park, J. K. and Cho, K. R. (2005). Experimental study of UWB antenna in the time domain. *Microwave and Optical Technology Letters.* 47:6, 554–558.

Lee, W. S., Kim, D. Z. and Yu, J. W. (2006). Wideband crossed planar monopole antenna with the band-notched characteristic. *Microwave and Optical Technology Letters.* 48:3, 543–545.

Li, P., Liang, J. and Chen, X. (2006). Study of printed elliptical/circular slot antenna for ultrawideband applications. *IEEE Transactions on Antennas and Propagation.* 54:6, 1670–1675.

Li, X., Bond, E. J., Van Veen, B. D. and Hagness, S. C. (2005), An overview of ultra-wideband microwave imaging via space-time beamforming for early-stage breast-cancer detection. *IEEE AP-S Magazine*. 47:1, 19–34.

Liang, J., Chiau, C. C., Chen, X. and Parini, C. G. (2007). Study of a printed circular disc monopole antenna for UWB systems. *IEEE Transactions on Antennas and Propagation*. 53:11, 3500–3504.

Liang, J., Guo, L., Chiau, C. C., Chen, X. and Parini, C. G. (2005). Study of CPW-fed circular disc monopole antenna for ultra wideband applications. *IEE Proceedings on Microwaves, Antennas and Propagation*. pp. 520–526.

Licul, S. and Davis, W. A. (2003). A comparison between a cavity backed spiral and a resonant monopole — An ultra-wideband consideration. *IEEE Antennas and Propagation Society International Symposium*. 4, 486–489.

Ling, C. and Li, S. (2000). Chaotic Spreading sequences with multiple access performance better than random sequences. *IEEE Transactions on CAS-I*. 47, 394–397.

Ling, C. and Sun, S. (1998). Chaotic frequency hopping sequences. *IEEE Transactions on Communications*. 46, 1433–1437.

Ling, C. and Wu, X. (2001). Design and implementation of an FPGA-based generator for chaotic frequency hopping sequences. *IEEE Transactions on CAS-I*. 48, 521–532.

Liu, W. C. (2004). Broadband dual-frequency meandered CPW-fed monopole antenna. *Electronic Letters*. 40, 642–643.

Liu, W. C. and Kao, P. C. (2005). CPW-fed triangular monopole antenna for ultra-wideband operation. *Microwave and Optical Technology Letters*. 47:6, 580–582.

Lo, Y. T. (1953). A note on the cylindrical antenna of noncircular cross section. *Journal of Applied Physics*. 24, 1338.

Lu, G., von der Mark, S., Korisch, I., Greenstein, L. J. and Spasojevic, P. (2004). Diamond and rounded diamond antennas for ultrawide-band communications. *IEEE Antennas and Wireless Propagation Letters*. 3:1, 249–252.

Ma, T. G. and Jeng, S. K. (2005). Planar miniature tapered-slot-fed annular slot antennas for ultrawide-band radios. *IEEE Transactions on Antennas and Propagation*. 53:3, 1194–1202.

Mao, S. G., Yeh, J. C. and Chen, S. L. (2009). Ultrawideband circularly polarized spiral antenna using integrated balun with application to time-domain target detection. *IEEE Transactions on Antennas and Propagation*. 57:7, 1914–1920.

Mayes, P. E. (1992). Frequency-independent antennas and broad-band derivates thereof. *IEEE Proceedings*. 80, 103–112.

McNamara, D., Baker, D. and Botha, L. (1984). Some design considerations for biconical antennas. *Antennas and Propagation Society International Symposium*. 22, 173–176.

Merli, F., Zurcher, J. F., Freni, A. and Skrivervik, A. K. (2009). Analysis, design and realization of a novel directive ultra-wideband antenna. *IEEE Transactions on Antennas and Propagation*. 57:11, 3458–3466.

Noguchi, K., Betsudan, S., Katagi, T. and Mizusawa, M. (2007). A compact broad-band helical antenna with two-wire helix. *IEEE Transactions on Antennas and Propagation*. 51:9, 2176–2181.

Okamoto, Y. and Hirose, A. (2008). Wideband adaptive antenna using selective feeding and stagger tuning. *Electronics Letters*. 44:19, 1116–1117.

Pendergrass, M. (2003). Time domain corporation. Time domain supporting text for 802.15.3 alternate Physical Layer Proposal. IEEE 802.15-03/144r1.

Pozar, D. M. (2005). *Microwave Engineering*. John Wiley & Sons, Danvers.

Qiu, R. C., Liu, H. and Shen, X. (2005). Ultra-wideband for multiple access communications. *IEEE Commun Magazine*. 43, 80–87.

Qiu, J., Du, Z., Lu, J. and Gong, K. (2005). A case study to improve the impedance bandwidth of a planar monopole. *Microwave and Optical Technology Letters*. 45:2, 124–126.

Rambabu, K., Thiart, H. A., Bornemann, J. and Yu, S. Y. (2006). Ultrawideband printed-circuit antenna. *IEEE Transactions of Antennas Propagations*. 54:12, 3908–3911.

Rudge, A. W., Milne, A. D. and Knight, P. (1986). *The Handbook of Antenna Design*. IEE. London.

Schantz, H. (2005). *The Art and Science of Ultrawideband Antennas*. Artech House, Noorwood, MA.

Schantz, H. G. and Fullerton, L. (2001). The diamond dipole: A gaussian impulse antenna. *IEEE Antennas and Propagation Society International Symposium*. 4, 100–103.

Sheng, H., Orlik, P., Haimovich, A. M., Cimini, L. J. and Zhang, Jr., J. (2003). On the spectral and power requirements for ultra-wideband transmission. *Proceedings of the IEEE International Conference on Communications*. 1, 738–742.

Schaubert, D., Kasturi, S., Elsallal, M. W. and Cappellen, W. V (2006). Wide bandwidth Vivaldi antenna arrays — some recent developments. *Proceedings of EuCAP 2006*. EuCAP, Nice. p. 117.1

See, T. S. P. and Chen, Z. N. (2008). A small UWB antenna for wireless USB. *IEEE APS-08*. Charleston, USA. pp. 198–203.

Simon, M. K., Omura, J. K., Scholtz, R. A. and Levitt, B. K. (1994). *Spread Spectrum Communications Handbook*. revised ed. McGraw-Hill, New York.

SkyCross (2009), 3.1–10GHz Ultra-wideband antenna, http://www.skycross.com/Products/PDFs/SMT-3TO10M-A.pdf.

Soergel, W., Waldschmidt, C. and Wiesbeck, W. (2003). Antenna characterization for ultra-wideband communications. *International Workshop on Ultrawideband Systems (IWUWBS)*. Oulu, Finland.

Stutzman, W. L. (1998). *Antenna Theory and Design*. John Wiley & Sons, New York.

Su, S. W., Wong, K. L. and Tang, C. L. (2004). Ultra-wideband square planar monopole antenna for IEEE 802.16a operation in the 2–11 GHz band. *Microwave and Optical Technology Letters*. 42:6, 463–466.

Suh, S. Y., Stutzman, W. L. and Davis, W. A. (2004). A new ultrawideband printed monopole antenna: The Planar Inverted Cone Antenna. (PICA). *IEEE Transactions on Antennas and Propagation*. 52:5, 1361–1365.

Sze, J. Y. and Shiu, J. Y. (2008). Design of band-notched ultrawideband square aperture antenna with a hat-shaped back-patch. *IEEE Transactions on Antennas and Propagation*. 56:10, 3311–3314.

Time Domain (2003), Manual of PulseON 200™ UWB Evaluation Kit.

Valderas, D., Meléndez, J. and Sancho, I. (2005). Some design criteria for UWB planar monopole antennas: Application to a slotted rectangular monopole. *Microwave and Optical Technology Letters*. 46:1, 6–11.

Valderas, D., Cendoya, I., Berenguer, R. and Sancho, I. (2006a). A method to optimize the bandwidth of UWB planar monopole antennas. *Microwave and Optical Technology Letters.* 48:2, 155–159.

Valderas, D., Legarda, J., Gutiérrez, I. and Sancho, J. I. (2006b). Design of UWB folded-plate monopole antennas based on TLM. *IEEE Transactions on Antennas and Propagation.* 54:6, 1676–1687.

Valderas, D., Sedano, B., García-Alonso A. and Sancho, J. I. (2007a). Synthesis of TLM-based UWB planar monopole impedance bandwidth. *IEEE Transactions on Antennas and Propagation.* 55:10, 2874–2879.

Valderas, D., de No, J., Meléndez, J. and Sancho, I. (2007b). Design of omnidirectional broadband metal-plate monopole antenna. *Microwave and Optical Technology Letters.* 49:2, 375–379.

Valderas, D., Álvarez, R., Meléndez, J., Gurutzeaga, I., Legarda, J. and Sancho, J. I. (2008). UWB staircase-profile printed monopole design. *IEEE Antennas and Wireless Propagation Letters.* 7, 255–259.

van Cappellen, W. A., de Jongh, R. V. and Ligthart, L. P. (2000). Potentials of ultra-short-pulse time-domain scattering measurements. *IEEE Antennas and Propagation Magazine.* 42:4, 35–45.

Vendelin, G. D., Pavio, A. M. and Rohde, U. L. (1990). *Microwave Circuit Design using Linear and Nonlinear Techniques.* Wiley Interscience, New York.

Wang, S., Guo, L., Chen, X., Parini, C. G. and McCormick, J. (2008). Design of compact broadband TSA arrays by using element mutual coupling. *Electronics Letters.* 44:18, 1049–1051.

Win, M. Z., Dadari, D., Molisch, A. F., Wiesbeck, W. and Zhang, J. (2009). History and applications of UWB. *Proc. IEEE* 97. pp. 198–204.

Win, M. Z. and Scholtz, R. A. (1998). Impulse radio: How it works. *IEEE Communcations Letters.* 2, 36–38.

Wong, K. L., Tseng, T. C. and Teng, P. L. (2004a). Low-profile ultra-wideband antenna for mobile phone applications. *Microwave and Optical Technology Letters.* 43:1, 7–9.

Wong, K. L., Chi, Y. W. and Wu, C. H. (2004b). Wideband tri-plate monopole antenna. *Electronics Letters.* 40:24, 1517–1519.

Wong, K. L., Chou, L. C. and Chen, H. T. (2004c). Ultra-wideband metal-plate monopole antenna for laptop application. *Microwave and Optical Technology Letters.* 43:5, 384–386.

Wong, K. L., Su, S. W. and Tang, C. L. (2005a). Broadband omnidirectional metal-plate monopole antenna. *IEEE Transactions on Antennas and Propagation.* 53:1, 581–583.

Wong, K. L., Wu, C. H. and Chang, F. S. (2005b). A compact wideband omnidirectional cross-plate monopole antenna. *Microwave and Optical Technology Letters.* 44:6, 492–494.

Yang, L. and Giannakis, G. B. (2004). Ultra-wideband communications: An idea whose time has come. *IEEE Signal Processing Magazine.* 21, 26–54.

Yazdandoost, K. Y. and Kohno, R. (2004). Ultra wideband antenna. *IEEE Communications Magazine.* 42:6, *S29–S32.*

Zhang, J., Orlik, P. V., Sahiloglu, Z., Molisch, A. F. and Kinney, P. (2009). UWB systems for wireless sensor networks. *Proc. IEEE* 97. 313–331.

Zentner, R., Bartolic, J. and Zentner, E. (2003). Broadband matching of Stacked Patch antennas using a single line-transformer technique. *Microwave and Optical Technology Letters*. 39:3, 178–183.

Index